# Introduction to Particle Technology

# Introduction to Particle Technology

**Martin Rhodes**
*Monash University, Australia*

JOHN WILEY & SONS
Chichester • New York • Weinheim • Brisbane • Singapore • Toronto

*Other Wiley Editorial Offices*

John Wiley & Sons, Inc., 605 Third Avenue,
New York, NY 10158-0012, USA

WILEY-VCH Verlag GmbH, Pappelallee 3,
D-69469 Weinheim, Germany

Jacaranda Wiley Ltd, 33 Park Road, Milton,
Queensland 4064, Australia

John Wiley & Sons (Asia) Pte Ltd, 2 Clementi Loop #02-01
Jin Xing Distripark, Singapore 0512

John Wiley & Sons (Canada) Ltd, 22 Worcester Road,
Rexdale, Ontario M9W 1L1, Canada

*Library of Congress Cataloguing in Publication Data*

Rhodes, M. J. (Martin J.)
    Introduction to particle technology / Martin Rhodes.
       p.  cm.
    Includes bibliographical references and index.
    ISBN 0-471-98482-5. — ISBN 0-471-98483-3
    1. Particles.  I. Title.
TP156.P3R.48  1998
620´.43 — dc21                                    98-7737
                                                  CIP

*British Library Cataloguing in Publication Data*

A catalogue record for this book is available from the British Library

ISBN 0 471 98482 5 (Cased)      0 471 98483 3 (paperback)

Typeset in 10/12pt Palatino by Keytec Typesetting Ltd, Bridport, Dorset
Printed and bound in Great Britain by Biddles Ltd, Guildford and King's Lynn.
This book is printed on acid-free paper responsibly manufactured from sustainable
forestry, in which at least two trees are planted for each one used in paper production.

# Contents

# Preface

## Particle Technology

Particle technology is a term used to refer to the science and technology related to the handling and processing of particles and powders. Particle technology is also often described as powder technology, particle science and powder science. Powders and particles are commonly referred to as bulk solids, particulate solids and granular solids. Today particle technology includes the study of liquid drops, emulsions and bubbles as well as solid particles. In this book only solid particles are covered and the terms particles, powder and particulate solids will be used interchangeably.

The discipline of particle technology now includes topics as diverse as the formation of aerosols and the design of bucket elevators, crystallisation and pneumatic transport, slurry filtration and silo design. A knowledge of particle technology may be used in the oil industry to design the catalytic cracking reactor which produces gasoline from oil or it may be used in forensic science to link the accused with the scene of the crime. Ignorance of particle technology may result in lost production, poor product quality, risk to health, dust explosion or storage silo collapse.

## Objective

The objective of this textbook is to introduce the subject of particle technology to students studying degree courses in disciplines requiring knowledge of the processing and handling of particles and powders. Although the primary target readership is amongst students of chemical engineering, the material included should form the basis of courses on particle technology for students studying other disciplines including mechanical engineering, civil engineering, applied chemistry, pharmaceutics, metallurgy and minerals engineering.

A number of key topics in particle technology are studied giving the fundamental science involved and linking this, wherever possible, to industrial practice. The coverage of each topic is intended to be exemplary rather than exhaustive. This is not intended to be a text on unit operations in powder technology for chemical engineers. Readers wishing to know more about the industrial practice and equipment for handling and processing are referred to the various handbooks of powder technology which are available.

The topics included have been selected to give coverage of broad areas within particle technology: characterisation (size analysis), processing (fluidized beds, granulation), particle formation (granulation, size reduction), fluid-particle-separation (filtration, settling, gas cyclones), safety (dust explosions), transport (pneumatic transport and standpipes). The health hazards of fine particles or dusts are not covered. This is not to suggest in any way that this topic is less important than others. It is omitted because of a lack of space and because the health hazards associated with dusts are dealt with competently in the many texts on Industrial or Occupational Hygiene which are now available. Students need to be aware however, that even chemically inert dusts or "nuisance dust" can be a major health hazard. Particularly where products contain a significant proportion of particles under 10 μm and where there is a possibility of the material becoming airborne during handling and processing. The engineering approach to the health hazard of fine powders should be strategic wherever possible; aiming to reduce dustiness by agglomeration, to design equipment for containment of material and to minimise exposure of workers.

The topics included demonstrate how the behaviour of powders is often quite different from the behaviour of liquids and gases. Behaviour of particulate solids may be surprising and often counter-intuitive when intuition is based on our experience with fluids. The following are examples of this kind of behaviour:

When a steel ball is placed at the bottom of a container of sand and the container is vibrated in a vertical plane, the steel ball will rise to the surface.

A steel ball resting on the surface of a bed of sand will sink swiftly if air is passed upward through the sand causing it to become fluidized.

Stirring a mixture of two free-flowing powders of different sizes may result in segregation rather than improved mixture quality.

Engineers and scientist are used to dealing with liquids and gases whose properties can be readily measured, tabulated and even calculated. The boiling point of pure benzene at one atmosphere pressure can be safely relied upon to remain at 80.1°C. The viscosity of water at 20°C can be confidently predicted to be 0.001 Pas. The thermal conductivity of copper at 100°C is 377 W/m.K. With particulate solids, the picture is quite different. The flow properties of sodium bicarbonate powder, for example, depends not only on the particle size distribution, the particle shape and surface properties, but also on the humidity of the atmosphere and the state of compaction of the powder. These variables are not easy to characterise and so their influence on the flow properties is difficult to predict with any confidence.

In the case of particulate solids it is almost always necessary to rely on performing appropriate measurements on the actual powder in question rather than relying on tabulated data. The measurements made are generally measurements of bulk properties, such as shear stress, bulk density, rather than measurements of fundamental properties such as particle size, shape and density. Although this is the present situation, in the not too distant future, we will be able to rely on sophisticated computer models for simulation of particulate systems. Mathematical modelling of particulate solids behaviour is a rapidly developing area of research around the world, and with increased computing power and better visualisation software, we will soon be able to link fundamental particle properties directly to bulk powder behaviour. It will even be possible to predict, from first principles, the influence of the presence of gases and liquids within the powder or to incorporate chemical reaction.

Particle technology is a fertile area for research. Many phenomena are still unexplained and design procedures rely heavily on past experience rather than on fundamental understanding. This situation presents exciting challenges to researchers from a wide range of scientific and engineering disciplines around the world. Many research groups have web sites which are interesting and informative at levels ranging from primary schools to serious researchers. Students are encouraged to visit these sites to find out more about particle technology. Our own web site at Monash University can be accessed via the Chemical Engineering Department web page at
http://www.eng.monash.edu.au/chemeng/

**Martin Rhodes**,
Mount Eliza, May 1998

# Acknowledgements

I would like to thank John Davidson, Emeritus Professor at Cambridge University for providing a selection of questions for inclusion in the chapters on fluidization and single particles in fluids. Thanks are also due to Peter Uhlherr, currently Honorary Senior Research Fellow at Monash University, for his suggestions of material to be included in the chapters on single particles, multiple particles and packed beds and for supplying worked examples and questions for these chapters.

I am grateful to my research workers Adam Forsyth Art Looi, Belinda Mathers, Ming Mao, Loretta McLaughlin, John Sanderson and Shan Wang for their help in reading chapters before the submission of the manuscript, preparation of photographs and proof reading.

Finally, I would like to acknowledge the part played in stimulating my interest in particles and powders by colleagues Derek Geldart, Norman Harnby, Arthur Hawkins, Nayland Stanley-Wood, Lado Svarovsky and John Williams, in the School of Powder Technology at my previous institution, Bradford University in England.

# 1

# Single Particles in a Fluid

This chapter deals with the motion of single solid particles in fluids. The objective here is to develop an understanding of the forces resisting the motion of any such particle and provide methods for the estimation of the steady velocity of the particle relative to the fluid. The subject matter of the chapter will be used in subsequent chapters on the behaviour of suspensions of particles in a fluid, fluidization, gas cyclones and pneumatic transport.

## 1.1 MOTION OF SOLID PARTICLES IN A FLUID

The drag force resisting very slow steady relative motion (creeping motion) between a rigid sphere of diameter $x$ and a fluid of infinite extent, of viscosity $\mu$ is composed of two components (Stokes, 1851):

$$\text{a pressure drag force, } F_{\mathrm{p}} = 2\pi x \mu U \tag{1.1}$$

$$\text{a shear stress drag force, } F_{\mathrm{s}} = \pi x \mu U \tag{1.2}$$

$$\text{Total drag force resisting motion, } F_{\mathrm{D}} = 3\pi x \mu U \tag{1.3}$$

where $U$ is the relative velocity.

This is known as Stokes' law. Experimentally, Stokes' law is found to hold almost exactly for single particle Reynolds number, $Re_{\mathrm{p}} \leqslant 0.1$, within 1% for $Re_{\mathrm{p}} \leqslant 0.3$, within 3% for $Re_{\mathrm{p}} \leqslant 0.5$ and within 9% for $Re_{\mathrm{p}} \leqslant 1.0$, where the single particle Reynolds number is defined in Equation (1.4).

$$\text{Single particle Reynolds number, } Re_{\mathrm{p}} = x U \rho_{\mathrm{f}}/\mu \tag{1.4}$$

$$\text{A drag coefficient, } C_{\mathrm{D}} \text{ is defined as } C_{\mathrm{D}} = R'/(\tfrac{1}{2}\rho_{\mathrm{f}} U^2) \tag{1.5}$$

where $R'$ is the force per unit projected area of the particle.

$$\text{Thus, for a sphere: } R' = F_D \bigg/ \left( \frac{\pi x^2}{4} \right) \tag{1.6}$$

and, Stokes' law, in terms of this drag coefficient, becomes:

$$C_D = 24/Re_p \tag{1.7}$$

At higher relative velocities, the inertia of the fluid begins to dominate (the fluid must accelerate out of the way of the particle). Analytical solution of the Navier–Stokes equations is not possible under these conditions. However, experiments give the relationship between the drag coefficient and the particle Reynolds number in the form of the so-called standard drag curve (Figure 1.1). Four regions are identified; the Stokes' law region, the Newton's law region in which drag coefficient is independent of Reynolds number, an intermediate region between the Stokes and Newton regions and the boundary layer separation region. The Reynolds number ranges and drag coefficient correlations for these regions are given in Table 1.1.

The expression given for $C_D$ in the intermediate region in Table 1.1 is that of Dallavalle (1948). An alternative is that of Schiller and Naumann (1933) (Equa-

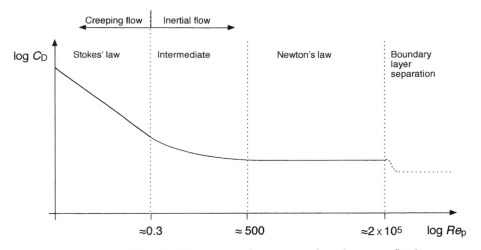

**Figure 1.1**   Standard drag curve for motion of a sphere in a fluid

**Table 1.1**   Reynolds number ranges for single particle drag coefficient correlations

| Region | Stokes | Intermediate | Newton's Law |
|---|---|---|---|
| $Re_p$ range | $<0.3$ | $0.3 < Re_p < 500$ | $500 < Re_p > 2 \times 10^5$ |
| $C_D$ | $24/Re_p$ | $\approx 24/Re_p + 0.44$ | $\approx 0.44$ |

tion (1.8)), which fits the data with an accuracy of around $\pm 7\%$ in the intermediate range.

$$C_D = \frac{24}{Re_p}(1 + 0.15 Re_p^{0.687}) \tag{1.8}$$

## 1.2   PARTICLES FALLING UNDER GRAVITY THROUGH A FLUID

The relative motion under gravity of particles in a fluid is of particular interest. In general, the forces of buoyancy, drag and gravity act on the particle:

$$\text{gravity} - \text{buoyancy} - \text{drag} = \text{acceleration force} \tag{1.9}$$

A particle falling from rest in a fluid will initially experience a high acceleration as the shear stress drag, which increases with relative velocity, will be small. As the particle accelerates the drag force increases, causing the acceleration to reduce. Eventually a force balance is achieved when the acceleration is zero and a maximum or terminal relative velocity is reached. This is known as the single particle terminal velocity.

For a spherical particle, Equation (1.9) becomes

$$\frac{\pi x^3}{6}\rho_p g - \frac{\pi x^3}{6}\rho_f g - R'\frac{\pi x^2}{4} = 0 \tag{1.10}$$

Combining Equation (1.10) with Equation (1.5),

$$\frac{\pi x^3}{6}(\rho_p - \rho_f)g - C_D\tfrac{1}{2}\rho_f U_T^2 \frac{\pi x^2}{4} = 0 \tag{1.11}$$

where $U_T$ is the single particle terminal velocity. Equation (1.11) gives the following expression for the drag coefficient under terminal velocity conditions:

$$C_D = \frac{4}{3}\frac{gx}{U_T^2}\left(\frac{(\rho_p - \rho_f)}{\rho_f}\right) \tag{1.12}$$

Thus in the Stokes' law region, with $C_D = 24/Re_p$, the single particle terminal velocity is given by

$$U_T = \frac{x^2(\rho_p - \rho_f)g}{18\mu} \tag{1.13}$$

Note that in the Stokes' law region the terminal velocity is proportional to the square of the particle diameter.

In the Newton's law region, with $C_D = 0.44$, the terminal velocity is given by

$$U_T = 1.74 \left( \frac{x(\rho_p - \rho_f)g}{\rho_f} \right)^{1/2} \tag{1.14}$$

Note that in this region the terminal velocity is independent of the fluid viscosity and proportional to the square root of the particle diameter.

In the intermediate region no explicit expression for $U_T$ can be found. However, in this region, the variation of terminal velocity with particle and fluid properties is approximately described by the following:

$$U_T \propto x^{1.1}, \ (\rho_p - \rho_f)^{0.7}, \ \rho_f^{-0.29}, \ \mu^{-0.43}$$

Generally, when calculating the terminal velocity for a given particle or the particle diameter for a given velocity, it is not known which region of operation is relevant. One way around this is to formulate the dimensionless groups, $C_D Re_p^2$ and $C_D/Re_p$:

- *To calculate $U_T$, for a given size x.* Calculate the group

$$C_D Re_p^2 = \frac{4}{3} \frac{x^3 \rho_f (\rho_p - \rho_f)g}{\mu^2} \tag{1.15}$$

which is independent of $U_T$
(Note that $C_D Re_p^2 = \frac{4}{3} Ar$, where $Ar$ is the Archimedes number)

For given particle and fluid properties, $C_D Re_p^2$ is a constant and will therefore produce a straight line of slope $-2$ if plotted on the logarithmic coordinates ($\log C_D$ versus $\log Re_p$) of the standard drag curve. The intersection of this straight line with the drag curve gives the value of $Re_p$ and hence $U_T$ (Figure 1.2).

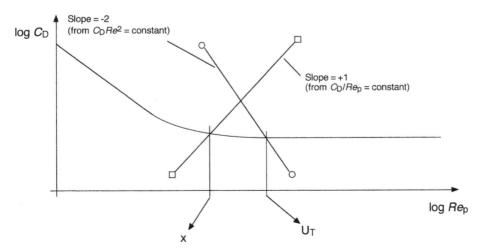

**Figure 1.2** Method for estimating terminal velocity for a given size of particle and vice versa (Note: $Re_p$ is based on the equivalent volume sphere diameter, $x_v$

- *To calculate size x, for a given $U_T$.* Calculate the group

$$\frac{C_D}{Re_p} = \frac{4}{3} \frac{g\mu(\rho_p - \rho_f)}{U_T^3 \rho_f^2}$$ (1.16)

which is independent of particle size $x$.

For a given terminal velocity, particle density and fluid properties, $C_D/Re_p$ is constant and will produce a straight line of slope $+1$ if plotted on the logarithmic coordinates ($\log C_D$ versus $\log Re_p$) of the standard drag curve. The intersection of this straight line with the drag curve gives the value of $Re_p$ and hence, $x$ (Figure 1.2).

## 1.3 NON-SPHERICAL PARTICLES

The effect of the shape of non-spherical particles on their drag coefficient is described by their sphericity, the ratio of the surface area of a sphere of volume equal to that of the particle to the surface area of the particle. For example, a cube of side one unit has a volume of 1 (cubic units) and a surface area of 6 (square units). A sphere of the same volume has a diameter, $x_v$, of 1.24 units. The surface area of a sphere of diameter 1.24 units is 4.836 cubic units. The sphericity of a cube is therefore 0.806 ($= 4.836/6$).

Shape affects drag coefficient far more in the intermediate and Newton's law regions than in the Stokes' law region. It is interesting to note that in the Stokes' law region particles fall with their longest surface nearly parallel to the direction of motion, whereas, in the Newton's law region particles present their maximum area to the oncoming fluid.

For non-spherical particles the particle Reynolds number is based on the equal-volume sphere diameter, i.e. the diameter of the sphere having the same volume as that of the particle. Figure 1.3 shows drag curves for particles of different sphericities.

Small particles in gases and all common particles in liquids quickly accelerate to their terminal velocity. As an example, a 100 μm particle falling from rest in water requires 1.5 ms to reach its terminal velocity of 2 mm/s. Table 1.2 gives some interesting comparisons of terminal velocities, acceleration times and distances for sand particles falling from rest in air.

## 1.4 EFFECT OF BOUNDARIES ON TERMINAL VELOCITY

When a particle is falling through a fluid in the presence of a solid boundary the terminal velocity reached by the particle is less than that for an infinite fluid. In the case of a particle falling along the axis of a vertical pipe this is described by a wall factor, $f_w$, the ratio of the velocity in the pipe, $U_{TD}$, to the velocity in an infinite fluid, $U_{T\infty}$.

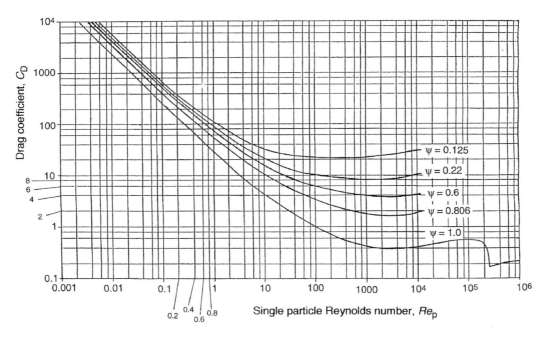

**Figure 1.3**   Drag coefficient $C_D$ versus Reynolds number $Re_p$ for particles of sphericity $\psi$ ranging form 0.125 to 1.0 (note $Re_p$ uses the equivalent volume diameter)

**Table 1.2**   Sand particles falling from rest in air (particle density, 2600 kg/m³)

| Size | Time to reach 99% of $U_T$ (s) | $U_T$ (m/s) | Distance travelled in this time (m) |
| --- | --- | --- | --- |
| 30 μm | 0.033 | 0.07 | 0.001 85 |
| 3 mm | 3.5 | 14 | 35 |
| 3 cm | 11.9 | 44 | 453 |

Examples of correlations for $f_w$ covering different ranges of Reynolds number and ratio of particle diameter to pipe diameter $(x/D)$ are those of Faxen (1923), Munroe (1888–89) and Francis (1933).

$$\text{Faxen:} \quad f_w = 1 - 2.1\left(\frac{x}{D}\right) \qquad Re_p \leqslant 0.3;\ x/D \leqslant 0.1 \qquad (1.17)$$

$$\text{Munroe:} \quad f_w = 1 - \left(\frac{x}{D}\right)^{1.5} \qquad 10^3 \leqslant Re_p \leqslant 10^4;\ 0.1 \leqslant x/D \leqslant 0.8 \qquad (1.18)$$

$$\text{Francis:} \quad f_w = \left(1 - \frac{x}{D}\right)^{2.25} \qquad Re_p \leqslant 0.3;\ x/D \leqslant 0.97 \qquad (1.19)$$

## 1.5 FURTHER READING

For further details on the motion of single particles in fluids (accelerating motion, added mass, bubbles and drops, non-Newtonian fluids) the reader is referred to Coulson and Richardson (1991), Clift *et al.* (1978) and Chhabra (1993).

## 1.6 WORKED EXAMPLES

### WORKED EXAMPLE 1.1

Calculate the upper limit of particle diameter $x_{max}$ as a function of particle density $\rho_p$ for gravity sedimentation in the Stokes' law regime. Plot the results as $x_{max}$ versus $\rho_p$ over the range $0 \leqslant \rho_p \leqslant 8000$ kg/m$^3$ for settling in water and in air at ambient conditions. Assume that the particles are spherical and that Stokes' law holds for $Re_p \leqslant 0.3$.

*Solution*

The upper limit of particle diameter in the Stokes' regime is governed by the upper limit of single particle Reynolds number:

$$Re_p = \frac{\rho_f x_{max} U}{\mu} = 0.3$$

In gravity sedimentation in the Stokes' regime particles accelerate rapidly to their terminal velocity. In the Stokes' regime the terminal velocity is given by Equation (1.13):

$$U_T = \frac{x^2(\rho_p - \rho_f)g}{18\mu}$$

Solving these two equations for $x_{max}$ we have

$$x_{max} = \left(0.3 \times \frac{18\mu^2}{g(\rho_p - \rho_f)\rho_f}\right)^{1/3}$$

$$= 0.82 \times \left(\frac{\mu^2}{(\rho_p - \rho_f)\rho_f}\right)^{1/3}$$

Thus, for air (density 1.2 kg/m$^3$ and viscosity $1.84 \times 10^{-5}$ Pa s):

$$x_{max} = 5.37 \times 10^{-4} \left(\frac{1}{(\rho_p - 1.2)}\right)^{1/3}$$

And for water (density 1000 kg/m$^3$ and viscosity 0.001 Pa s):

$$x_{max} = 8.19 \times 10^{-4} \left( \frac{1}{(\rho_p - 1000)} \right)^{1/3}$$

These equations for $x_{max}$ as a function are plotted in Figure 1.W1.1 for particle densities greater than and less than the fluid densities.

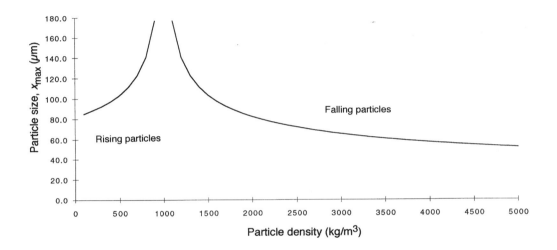

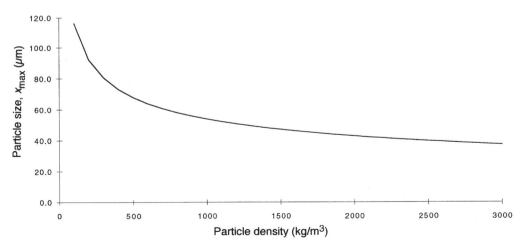

**Figure 1.W1.1**  (a) Limiting particle size for Stokes' law in water (Worked Example 1.1); (b): Limiting particle size for Stokes' law in air (Worked Example 1.1)

## WORKED EXAMPLE 1.2

A gravity separator for the removal from water of oil droplets (assumed to behave as rigid spheres) consists of a rectangular chamber containing inclined baffles as shown schematically in Figure 1.W2.1.

(a) Derive an expression for the ideal collection efficiency of this separator as a function of droplet size and properties, separator dimensions, fluid properties and fluid velocity (assumed uniform).

(b) Hence, calculate the percent change in collection efficiency when the throughput of water is increased by a factor of 1.2 and the density of the oil droplets changes from 750 to 850 kg/m$^3$.

### Solution

(a) Referring to Figure 1.W2.1, we will assume that all particles falling on to the top surface of a baffle will be collected. Therefore, any particle which can fall a distance $h$ or greater in the time required for it to travel the length of the separator will be collected. Let the corresponding minimum vertical droplet velocity be $U_{T_{min}}$.

Assuming uniform fluid velocity and negligible relative velocity between drops and fluid in the horizontal direction,

$$\text{drop residence time,} \quad t = L/U$$

$$\text{Then } U_{T_{min}} = hU/L$$

Assuming that the droplets are small enough for Stokes' law to apply and that the time and distance for acceleration to terminal velocity is negligible, then droplet velocity will be given by Equation (1.13):

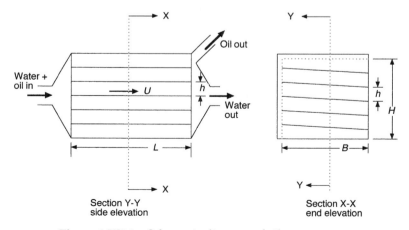

Section Y-Y
side elevation

Section X-X
end elevation

**Figure 1.W2.1**  Schematic diagram of oil–water separator

$$U_T = \frac{x^2(\rho_p - \rho_f)g}{18\mu}$$

This is the minimum velocity for drops to be collected whatever their original position between the baffles. Thus

$$U_{T_{min}} = \frac{x^2(\rho_p - \rho_f)g}{18\mu} = \frac{hU}{L}$$

Assuming that drops of all sizes are uniformly distributed over the vertical height of the separator, then drops falling a distance less than $h$ in the time required for them to travel the length of the separator will be fractionally collected depending on their original vertical position between two baffles. Thus for droplets falling at a velocity of $0.5U_{T_{min}}$ only 50% will be collected; i.e. only those drops originally in the lower half of the space between adjacent baffles. For drops falling at a velocity of $0.25U_{T_{min}}$ only 25% will be collected.

$$\text{Thus, efficiency of collection, } \eta = \frac{\text{actual } U_T \text{ for droplet}}{U_{T_{min}}}$$

$$\text{and so, } \eta = \left(\frac{x^2(\rho_p - \rho_f)g}{18\mu}\right) \Big/ \frac{hU}{L}$$

(b) Comparing collection efficiencies when the throughput of water is increased by a factor of 1.2 and the density of the oil droplets changes from 750 to 850 kg/m$^3$.

Let original and new conditions be denoted by subscripts 1 and 2 respectively.

Increasing throughput of water by a factor of 1.2 means that $U_2/U_1 = 1.2$. Therefore from the expression for collection efficiency derived above:

$$\frac{\eta_2}{\eta_1} = \left[\frac{\rho_{p2} - \rho_f}{U_2}\right] \Big/ \left[\frac{\rho_{p1} - \rho_f}{U_1}\right]$$

$$\frac{\eta_2}{\eta_1} = \left[\frac{850 - 1000}{750 - 1000}\right] \times \frac{1}{1.2} = 0.5$$

The decrease in collection efficiency is therefore 50%.

## WORKED EXAMPLE 1.3

A sphere of diameter 10 mm and density 7700 kg/m$^3$ falls under gravity at terminal conditions through a liquid of density 900 kg/m$^3$ in a tube of diameter 12 mm. The measured terminal velocity of the particle is 1.6 mm/s. Calculate the viscosity of the fluid. Verify that Stokes' law applies.

## Solution

To solve this problem, we first convert the measured terminal velocity to the equivalent velocity which would be achieved by the sphere in a fluid of infinite extent. Assuming Stokes' law we can determine the fluid viscosity. Finally we check the validity of Stokes' law.

Using the Francis wall factor expression (Equation (1.19)):

$$\frac{U_{T_\infty}}{U_{T_D}} = \frac{1}{(1 - x/D)^{2.25}} = 56.34$$

Thus, terminal velocity for the particle in a fluid of infinite extent,

$$U_{T_\infty} = U_{T_D} \times 56.34 = 0.0901 \text{ m/s}$$

Equating this value to the expression for $U_{T_\infty}$ in the Stokes' regime (Equation (1.13)):

$$U_{T_\infty} = \frac{(10 \times 10^{-3})^2 \times (7700 - 900) \times 9.81}{18\mu}$$

Hence, fluid viscosity, $\mu = 4.11 \text{ Pa s}$

Checking the validity of Stokes' law:

$$\text{Single particle Reynolds number, } Re_p = \frac{x \rho_f U}{\mu} = 0.197$$

$Re_p$ is less than 0.3 and so the assumption that Stokes' law holds is valid.

## WORKED EXAMPLE 1.4

A mixture of spherical particles of two materials A and B is to be separated using a rising stream of liquid. The size range of both materials is 15–40 μm. (a) Show that a complete separation is not possible using water as the liquid. The particle densities for materials A and B are 7700 and 2400 kg/m$^3$ respectively. (b) Which fluid property must be changed to achieve complete separation? Assume Stokes' law applies.

## Solution

(a) First, consider what happens to a single particle introduced into the centre of a pipe in which a fluid is flowing upwards at a velocity $U$ which is uniform across the pipe cross section. We will assume that the particle is small enough to consider the time and distance for its acceleration to terminal velocity to be negligible. Referring to Figure 1.W4.1(a), if the fluid velocity is greater than the terminal velocity of the particle $U_T$, then the particle will move upwards; if the fluid velocity is less than $U_T$ the particle will fall and if the fluid velocity is equal to $U_T$, the particle will remain at the same vertical position. In each case the velocity of the particle relative to the pipe wall is $(U - U_T)$. Now consider introducing two particles of different size and density having terminal velocities $U_{T_1}$ and $U_{T_2}$. Referring to Figure 1.W4.1(b), at low fluid velocities ($U < U_{T_2} < U_{T_1}$), both particles will fall. At high fluid velocities $U > U_{T_1} > U_{T_2}$), both particles will be carried

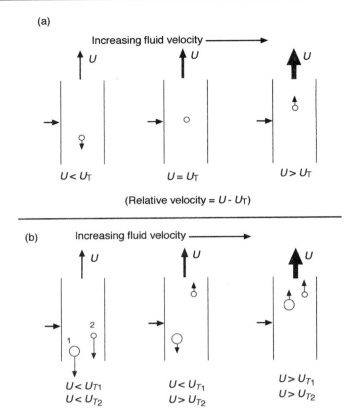

**Figure 1.W4.1**    Relative motion of particles in a moving fluid. (Worked Example 1.4)

upwards. At intermediate fluid velocities ($U_{T_1} > U > U_{T_2}$), particle 1 will fall and particle 2 will rise. Thus we have the basis of a separator according to particle size and density. From the analysis above we see that to be able to completely separate particles A and B, there must be no overlap between the ranges of terminal velocity for the particles; i.e. all sizes of the denser material A must have terminal velocities which are greater than all sizes of the less dense material B.

Assuming Stokes' law applies, Equation (1.13), with fluid density and viscosity 1000 kg/m$^3$ and 0.001 Pa s respectively, gives

$$U_T = 545x^2(\rho_p - 1000)$$

Based on this equation, the terminal velocities of the extreme sizes of particles A and B are:

| Size ($\mu$m) $\rightarrow$ | 15 | 40 |
|---|---|---|
| $U_{T_A}$ (mm/s) | 0.82 | 5.84 |
| $U_{T_B}$ (mm/s) | 0.17 | 1.22 |

We see that there is overlap of the ranges of terminal velocities. We can therefore select no fluid velocity which would completely separate particles A and B.

(b) Inspecting the expression for terminal velocity in the Stokes' regime (Equation (1.13)) we see that changing the fluid viscosity will have no effect on our ability to separate the particles, since change in viscosity will change the terminal velocities of all particles in the same proportion. However, changing the fluid density will have a different effect on particles of different density and this is the effect we are looking for. The critical condition for separation of particles A and B is when the terminal velocity of the smallest A particle is equal to the terminal velocity of the largest B particle.

$$U_{T_{B40}} = U_{T_{A15}}$$

Hence,

$$545 \times x_{40}^2 \times (2400 - \rho_f) = 545 \times x_{15}^2 \times (7700 - \rho_f)$$

From which, critical minimum fluid density $\rho_f = 1533 \text{ kg/m}^3$

## WORKED EXAMPLE 1.5

A sphere of density $2500 \text{ kg/m}^3$ falls freely under gravity in a fluid of density $700 \text{ kg/m}^3$ and viscosity $0.5 \times 10^{-3} \text{ Pa s}$. Given that the terminal velocity of the sphere is 0.15 m/s, calculate its diameter. What would be the edge length of a cube of the same material falling in the same fluid at the same terminal velocity?

### Solution

In this case we know the terminal velocity, $U_T$, and need to find the particle size $x$. Since we do not know which regime is appropriate, we must first calculate the dimensionless group $C_D/Re_p$ (Equation (1.16)):

$$\frac{C_D}{Re_p} = \frac{4}{3} \frac{g\mu(\rho_p - \rho_f)}{U_T^3 \rho_f^2}$$

hence,

$$\frac{C_D}{Re_p} = \frac{4}{3} \left[ \frac{9.81 \times (0.5 \times 10^{-3}) \times (2500 - 700)}{0.15^3 \times 700^2} \right]$$

$$\frac{C_D}{Re_p} = 7.12 \times 10^{-3}$$

This is the relationship between drag coefficent $C_D$ and single particle Reynolds number $Re_p$ for particles of density $2500 \text{ kg/m}^3$ having a terminal velocity of $0.15 \text{ m/s}$ in a fluid of density $700 \text{ kg/m}^3$ and viscosity $0.5 \times 10^{-3} \text{ Pa s}$. Since

$C_D/Re_p$ is a constant, this relationship will give a straight line of slope +1 when plotted on the log–log coordinates of the standard drag curve.

For plotting the relationship:

| $Re_p$ | $C_D$ |
|--------|-------|
| 100 | 0.712 |
| 1000 | 7.12 |
| 10 000 | 71.2 |

These values are plotted on the standard drag curves for particles of different sphericity (Figure 1.3). The result is shown in Figure 1.W5.1.

Where the plotted line intersects the standard drag curve for a sphere ($\psi = 1$), $Re_p = 130$.

The diameter of the sphere may be calculated from:

$$Re_p = 130 = \frac{\rho_f x_v U_T}{\mu}$$

Hence, sphere diameter $x_v = 619$ μm.

For a cube having the same terminal velocity under the same conditions, the same $C_D$ versus $Re_p$ relationship applies, only the standard drag curve is that for a cube ($\psi = 0.806$).

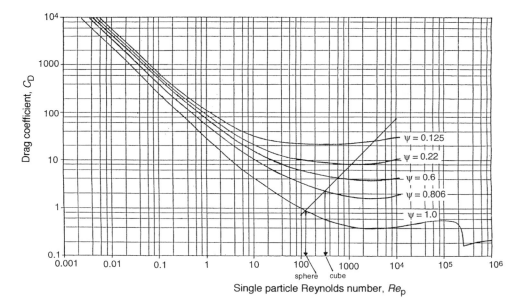

**Figure 1.W5.1** Drag coefficient $C_D$ versus Reynolds number, $Re_p$. Solution to Worked Example 1.5

Cube sphericity

For a cube of side 1 unit, the volume is 1 cubic unit and the surface area is 6 square units.

If $x_v$ is the diameter of a sphere having the same volume as the cube, then

$$\frac{\pi x_v^3}{6} = 1.0 \text{ which gives } x_v = 1.24 \text{ units}$$

Therefore, sphericity $\psi = \dfrac{\text{surface area of a sphere of volume equal to the particle}}{\text{surface area of the particle}}$

$$\psi = \frac{4.836}{6} = 0.806$$

At the intersection of this standard drag curve with the plotted line, $Re_p = 310$.

Recalling that the Reynolds number in this plot uses the equivalent volume sphere diameter,

$$x_v = \frac{310 \times (0.5 \times 10^{-3})}{0.15 \times 700} = 1.48 \times 10^{-3} \text{ m}$$

And so the volume of the particle is $\dfrac{\pi x_v^3}{6} = 1.66 \times 10^{-9} \text{ m}^3$

Giving a cube side length of $(1.66 \times 10^{-9})^{1/3} = 1.18 \times 10^{-3} \text{ m} (1.18 \text{ mm})$

### WORKED EXAMPLE 1.6

A particle of equivalent volume diameter 0.5 mm, density 2000 kg/m$^3$ and sphericity 0.6 falls freely under gravity in a fluid of density 1.6 kg/m$^3$ and viscosity $2 \times 10^{-5}$ Pa s. Estimate the terminal velocity reached by the particle.

### Solution

In this case we know the particle size and are required to determine its terminal velocity without knowing which regime is appropriate. The first step is, therefore, to calculate the dimensionless group $C_D Re_p^2$:

$$C_D Re_p^2 = \frac{4}{3} \frac{x^3 \rho_f (\rho_p - \rho_f) g}{\mu^2}$$

$$= \frac{4}{3} \left[ \frac{(0.5 \times 10^{-3})^3 \times 1.6 \times (2000 - 1.6) \times 9.81}{(2 \times 10^{-5})^2} \right]$$

$$= 13\,069$$

This is the relationship between drag coefficient $C_D$ and single particle Reynolds number $Re_p$ for particles of size 0.5 mm and density 2000 kg/m$^3$ falling in a fluid of density 1.6 kg/m$^3$ and viscosity $2 \times 10^{-5}$ Pa s. Since $C_D Re_p^2$ is a constant, this relationship will give a straight line of slope $-2$ when plotted on the log–log coordinates of the standard drag curve.

For plotting the relationship:

| $Re_p$ | $C_D$ |
|--------|-------|
| 10 | 130.7 |
| 100 | 1.307 |
| 1000 | 0.013 |

These values are plotted on the standard drag curves for particles of different sphericity (Figure 1.3). The result is shown in Figure 1.W6.1.

Where the plotted line intersects the standard drag curve for a sphericity of $0.6(\psi = 0.6)$, $Re_p = 40$.

The terminal velocity $U_T$ may be calculated from

$$Re_p = 40 = \frac{\rho_f x_v U_T}{\mu}$$

Hence, terminal velocity, $U_T = 1.0$ m/s.

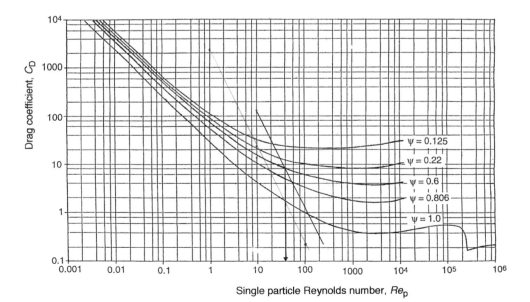

**Figure 1.W6.1** Drag coefficient $C_D$ versus Reynolds number $Re_p$. Solution to Worked Example 1.6

## EXERCISES

**1.1** The settling chamber, shown schematically in Figure 1.E1.1, is used as a primary separation device in the removal of dust particles of density 1500 kg/m³ from a gas of density 0.7 kg/m³ and viscosity $1.90 \times 10^{-5}$ Pa s.

(a) Assuming Stokes' law applies, show that the efficiency of collection of particles of size $x$ is given by the expression

$$\text{collection efficiency, } \eta_x = \frac{x^2 g(\rho_p - \rho_f)L}{18\mu HU}$$

where $U$ is the uniform gas velocity through the parallel-sided section of the chamber. State any other assumptions made.

(b) What is the upper limit of particle size for which Stokes' law applies?

(c) When the volumetric flow rate of gas is 0.9 m³/s, and the dimensions of the chamber are those shown in Figure 1.E1.1, determine the collection efficiency for spherical particles of diameter 30 μm. (Answer: (b) 57 μm; (c) (86%))

**1.2** A particle of equivalent sphere volume diameter 0.2 mm, density 2500 kg/m³ and sphericity 0.6 falls freely under gravity in a fluid of density 1.0 kg/m³ and viscosity $2 \times 10^{-5}$ Pa s. Estimate the terminal velocity reached by the particle. (Answer: 0.6 m/s)

**1.3** Spherical particles of density 2500 kg/m³ and in the size range 20–100 μm are fed continuously into a stream of water (density, 1000 kg/m³ and viscosity, 0.001 Pa s) flowing upwards in a vertical, large-diameter pipe. What maximum

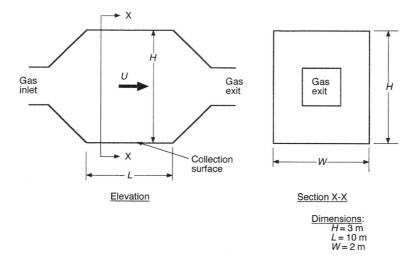

**Figure 1.E1.1** Schematic diagram of settling chamber

water velocity is required to ensure that no particles of diameter greater than 60 μm are carried upwards with water? (Answer: 2.9 mm/s)

**1.4** Spherical particles of density 2000 kg/m$^3$ and in the size range 20–100 μm are fed continuously into a stream of water (density, 1000 kg/m$^3$ and viscosity, 0.001 Pa s) flowing upwards in a vertical, large diameter pipe. What maximum water velocity is required to ensure that no particles of diameter greater than 50 μm are carried upwards with the water? (Answer: 1.4 mm/s)

**1.5** A particle of equivalent volume diameter 0.3 mm, density 2000 kg/m$^3$ and sphericity 0.6 falls freely under gravity in a fluid of density 1.2 kg/m$^3$ and viscosity 2 × 10$^{-5}$ Pa s. Estimate the terminal velocity reached by the particle. (Answer: 0.67 m/s)

**1.6** Assuming that a car is equivalent to a flat plate 1.5 m square, moving normal to the airstream, and with a drag coefficient, $C_D = 1.1$, calculate the power required for steady motion at 100 km/h on level ground. What is the Reynolds number? For air assume a density of 1.2 kg/m$^3$ and a viscosity of 1.71 × 10$^{-5}$ Pa s. (Cambridge University) (Answer: 31.9 kW; 2.95 × 10$^6$)

**1.7** A cricket ball is thrown with a Reynolds number such that the drag coefficient is 0.4 (Re ≈ 10$^5$).

(a)  Find the percentage change in velocity of the ball after 100 m horizontal flight in air.

(b)  With a higher Reynolds number and a new ball, the drag coefficient falls to 0.1. What is now the percentage change in velocity over 100 m horizontal flight?

(In both cases take the mass and diameter of the ball as 0.15 kg and 6.7 cm respectively and the density of air as 1.2 kg/m$^3$.) Readers unfamiliar with the game of cricket may substitute a baseball. (Cambridge University) (Answer: (a) 43.1%; (b) 13.1%)

**1.8** The resistance $F$ of a sphere of diameter $x$, due to its motion with velocity $u$ through a fluid of density $\rho$ and viscosity $\mu$, varies with Reynolds number ($Re = \rho u x / \mu$) as given below:

| $\log_{10} Re$ | 2.0 | 2.5 | 3.0 | 3.5 | 4.0 |
|---|---|---|---|---|---|
| $C_D = \dfrac{F}{\frac{1}{2}\rho u^2 (\pi x^2/4)}$ | 1.05 | 0.63 | 0.441 | 0.385 | 0.39 |

Find the mass of a sphere of 0.013 m diameter which falls with a steady velocity of 0.6 m/s in a large deep tank of water of density 1000 kg/m$^3$ and viscosity 0.0015 Pa s. (Cambridge University) (Answer: 0.0021 kg)

# 2

# Multiple Particle Systems

## 2.1 SETTLING OF A SUSPENSION OF PARTICLES

When many particles flow in a fluid in close proximity to each other the motion of each particle is influenced by the presence of the others. The simple analysis for the fluid particle interaction for a single particle is no longer valid but can be adapted to model the multiple particle system.

For a suspension of particles in a fluid, Stoke's law is assumed to apply but an effective suspension viscosity and effective average suspension density are used:

$$\text{effective viscosity, } \mu_e = \mu/f(\varepsilon) \tag{2.1}$$

$$\text{average suspension density, } \rho_{ave} = \varepsilon\rho_f + (1-\varepsilon)\rho_p \tag{2.2}$$

where $\varepsilon$ is the voidage or volume fraction occupied by the fluid. The effective viscosity of the suspension is seen to be equal to the fluid viscosity, $\mu$ modified by a function $f(\varepsilon)$ of the fluid volume fraction.

The drag coefficient for a single particle in the Stokes law region was shown in Chapter 1. to be given by $C_D = 24/Re_p$. Substituting the effective viscosity and average density for the suspension, Stokes law becomes

$$C_D = \frac{24}{Re_p} = \frac{24\mu_e}{U_{rel}\rho_{ave}x} \tag{2.3}$$

where, $C_D = R'/(\frac{1}{2}\rho_{ave}U_{rel}^2)$ and $U_{rel}$ is the relative velocity of the particle to the fluid.

Under terminal velocity conditions for a particle falling under gravity in a suspension, the force balance,

$$\text{drag force} = \text{weight} - \text{upthrust}$$

becomes

$$\left(\frac{\pi x^2}{4}\right)\frac{1}{2}\rho_{ave}U_{rel}^2 C_D = (\rho_p - \rho_{ave})\left(\frac{\pi x^3}{6}\right)g \tag{2.4}$$

giving

$$U_{rel} = (\rho_p - \rho_{ave}) \frac{x^2 g}{18\mu_e} \tag{2.5}$$

Substituting for average density, $\rho_{ave}$ and effective viscosity $\mu_e$ of the suspension, we obtain the following expression for the terminal falling velocity for a particle in a suspension, $U_{rel_T}$:

$$U_{rel_T} = (\rho_p - \rho_f) \frac{x^2 g}{18\mu} \varepsilon f(\varepsilon) \tag{2.6}$$

Comparing this with the expression for the terminal free fall velocity of a single particle in a fluid (Equation (1.13)), we find that

$$U_{rel_T} = U_T \varepsilon f(\varepsilon) \tag{2.7}$$

$U_{rel_T}$ is known as the particle settling velocity in the presence of other particles or the hindered settling velocity.

In the following analysis, it is assumed that the fluid and the particles are incompressible and that the volume flowrates, $Q_f$ and $Q_p$, of the fluid and the particles are constant.

We define $U_{fs}$ and $U_{ps}$ as the superficial velocities of the fluid and particles respectively:

$$\text{superficial fluid velocity, } U_{fs} = \frac{Q_f}{A} \tag{2.8}$$

$$\text{superficial particle velocity, } U_{ps} = \frac{Q_p}{A} \tag{2.9}$$

where $A$ is the vessel cross-sectional area.

Under isotropic conditions the flow areas occupied by the fluid and the particles are:

$$\text{flow area occupied by the fluid, } A_f = \varepsilon A \tag{2.10}$$

$$\text{flow area occupied by the particles, } A_p = (1 - \varepsilon)A \tag{2.11}$$

And so continuity gives:

$$\text{for the fluid:} \quad Q_f = U_{ps}A = U_f A \varepsilon \tag{2.12}$$

$$\text{for the particles:} \quad Q_p = U_{ps}A = U_p A(1 - \varepsilon) \tag{2.13}$$

hence the actual velocities of the fluid and the particles, $U_f$ and $U_p$ are given by:

$$\text{actual velocity of the fluid, } U_f = U_{fs}/\varepsilon \tag{2.14}$$

$$\text{actual velocity of the particles, } U_p = U_{ps}/(1 - \varepsilon) \tag{2.15}$$

## 2.2   BATCH SETTLING

### 2.2.1   Settling Flux as a Function of Suspension Concentration

When a batch of solids in suspension are allowed to settle, say in a measuring cylinder in the laboratory, there is no net flow through the vessel and so

$$Q_p + Q_f = 0 \qquad (2.16)$$

hence,

$$U_p(1 - \varepsilon) + U_f \varepsilon = 0 \qquad (2.17)$$

and,

$$U_f = -U_p \frac{(1 - \varepsilon)}{\varepsilon} \qquad (2.18)$$

In hindered settling under gravity the relative velocity between the particles and the fluid $(U_p - U_f)$ is $U_{rel_T}$. Thus using the expression for $U_{rel_T}$ found in Equation (2.7), we have

$$U_p - U_f = U_{rel_T} = U_T \varepsilon f(\varepsilon) \qquad (2.19)$$

Combining Equation (2.19) with Equation (2.18) gives the following expression for $U_p$, the hindered settling velocity of particles relative to the vessel wall in batch settling:

$$U_p = U_T \varepsilon^2 f(\varepsilon) \qquad (2.20)$$

The effective viscosity function, $f(\varepsilon)$, was shown theoretically by Einstein to be

$$f(\varepsilon) = \varepsilon^{2.5} \qquad (2.21)$$

for uniform spheres forming a suspension of solid volume fraction less than 0.1 $\{(1 - \varepsilon) \leqslant 0.1\}$.

Richardson and Zaki [1954] showed by experiment that for $Re_p < 0.3$ (under Stoke's law conditions where drag is independent of fluid density),

$$U_p = U_T \varepsilon^{4.65} \qquad \text{(giving } f(\varepsilon) = \varepsilon^{2.65}) \qquad (2.22)$$

and for $Re_p > 500$ (under Newton's law conditions where drag is independent of fluid viscosity)

$$U_p = U_T \varepsilon^{2.4} \qquad \text{(giving } f(\varepsilon) = \varepsilon^{0.4}) \qquad (2.23)$$

In general, the Richardson and Zaki relationship is given as

$$U_p = U_T \varepsilon^n \qquad (2.24)$$

Khan and Richardson [1989] recommend the use of the following correlation for the value of exponent $n$ over the entire range of Reynolds numbers:

$$\frac{4.8 - n}{n - 2.4} = 0.043 Ar^{0.57} \left[ 1 - 2.4 \left( \frac{x}{D} \right)^{0.27} \right] \qquad (2.25)$$

where Ar is the Archimedes number $[x^3 \rho_f (\rho_p - \rho_f) g / \mu^2]$ and $x$ is the particle diameter and $D$ is the vessel diameter.

Expressed as a volumetric solids settling flux, $U_{ps}$, Equation (2.24) becomes

$$U_{ps} = U_p (1 - \varepsilon) = U_T (1 - \varepsilon) \varepsilon^n \qquad (2.26)$$

or, dimensionless particle settling flux,

$$\frac{U_{ps}}{U_T} = (1 - \varepsilon) \varepsilon^n \qquad (2.27)$$

Taking first and second derivates of Equation (2.27) demonstrates that a plot of dimensionless particle settling flux versus suspension volumetric concentration, $1 - \varepsilon$ has a maximum at $\varepsilon = n/(n+1)$ and an inflection point at $\varepsilon = (n-1)/(n+1)$. The theoretical form of such a plot is therefore that shown in Figure (2.1).

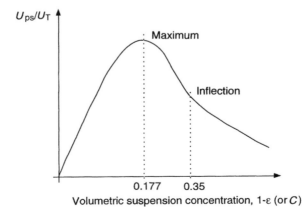

**Figure 2.1** Variation of dimensionless settling flux with suspension concentration, based on Equation 2.27 (For $Re_p < 0.3$; i.e. $n = 4.65$)

## 2.2.2  Sharp Interfaces in Sedimentation

Interfaces or discontinuities in concentration occur in the sedimentation or settling of particle suspensions.

In the remainder of this chapter, for convenience, the symbol $C$ will be used to represent the particle volume fraction $1 - \varepsilon$. Also for convenience the particle volume fraction will be called the concentration of the suspension.

Consider Figure 2.2, which shows the interface between a suspension of concentration $C_1$ containing particles settling at a velocity $U_{p_1}$ and a suspension of concentration $C_2$ containing particles settling at a velocity $U_{p_2}$.

The interface is falling at a velocity $U_{\text{int}}$. All velocities are measured relative to the vessel walls. Assuming incompressible fluid and particles, the mass balance over the interface gives

$$(U_{p_1} - U_{\text{int}})C_1 = (U_{p_2} - U_{\text{int}})C_2$$

hence

$$U_{\text{int}} = \frac{U_{p_1}C_1 - U_{p_2}C_2}{C_1 - C_2} \tag{2.28}$$

or, since $U_p C$ is particle volumetric flux, $U_{ps}$, then:

$$U_{\text{int}} = \frac{U_{ps_1} - U_{ps_2}}{C_1 - C_2} \tag{2.29}$$

where $U_{ps_1}$ and $U_{ps_2}$ are the particle volumetric fluxes in suspensions of concentration $C_1$ and $C_2$ respectively. Thus,

$$U_{\text{int}} = \frac{\Delta U_{ps}}{\Delta C} \tag{2.30}$$

and, in the limit as $\Delta C \to 0$, $U_{\text{int}} = \dfrac{dU_{ps}}{dC}$ \hfill (2.31)

Hence, on a flux plot (a plot of $U_{ps}$ versus concentration):

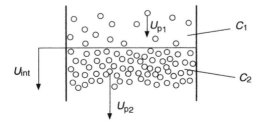

**Figure 2.2**  Concentration interface in sedimentation

(a)   The gradient of the curve at concentration $C$ is the velocity of a layer of suspension of this concentration.

(b)   The slope of a chord joining two points at concentrations $C_1$ and $C_2$ is the velocity of a discontinuity or interface between suspensions of these concentrations.

This is illustrated in Figure 2.3.

### 2.2.3   The Batch Settling Test

The simple batch settling test can supply all the information for the design of a thickener for separation of particles from a fluid. In this test a suspension of particles of known concentration is prepared in a measuring cylinder. The cylinder shaken to thoroughly mix the suspension and then placed upright to allow the suspension to settle. The positions of the interfaces which form are monitored in time. Two types of settling occur depending on the initial concentration of the suspension. The first type of settling is depicted in Figure 2.4 (Type 1 settling). Three zones of constant concentration are formed. These are zone A, clear liquid ($C = 0$); zone B, of concentration equal to the initial suspension concentration ($C_B$); and zone S, the sediment concentration ($C_S$). Figure 2.5 is a typical plot of the height of the interfaces AB, BS and AS with time for this type of settling. On this plot the slopes of the lines give the velocities of the interfaces. For example, interface AB descends at constant velocity, interface BS rises at constant velocity. The test ends when the descending AB meets the rising BS forming an interface between clear liquid and sediment (AS) which is stationary.

In the second type of settling (Type 2 settling), shown in Figure 2.6, a zone of variable concentration, zone E, is formed in addition to the zones of constant concentration (A, B and S). The suspension concentration within zone E varies

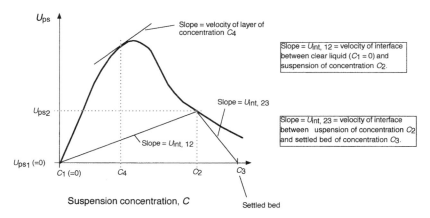

**Figure 2.3**   Determination of interface and layer velocities from a batch flux plot

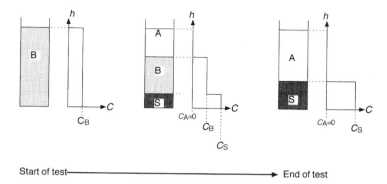

Start of test ————————————————————→ End of test

**Figure 2.4** Type 1 batch settling. Zones A, B and S are zones of constant concentration. Zone A is a clear liquid; zone B is a suspension of concentration equal to the initial suspension concentration; zone S is a suspension of settled bed or sediment concentration.

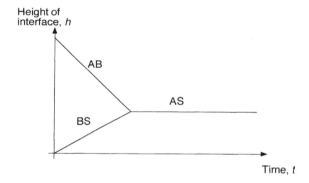

**Figure 2.5** Change in positions of interface AB, BS and AS with time in type 1 batch settling (e.g. AB is the interface between zone A and zone B; see Figure 2.4)

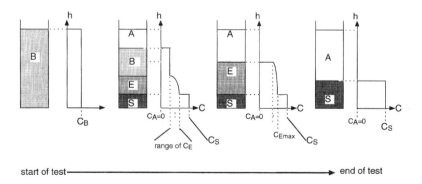

start of test ————————————————————→ end of test

**Figure 2.6** Type 2 batch settling. Zones A, B and S are zones of constant concentration. Zone A is clear liquid; zone B is a suspension of concentration equal to the initial suspension concentration; zone S is a suspension of settled bed concentration. Zone E is a zone of variable concentration with position. However, the minimum and maximum concentrations within this zone, $C_{Emin}$ and $C_{Emax}$, are constant. Figure 2.7 is a typical plot of the height of the interfaces AB, $BE_{min}$, $E_{max}S$ and AS with time for this type of settling.

The occurrence of type 1 or type 2 settling depends on the initial concentration of the suspension, $C_B$. To understand why a zone of variable concentration forms in one case but not in the other we must consider the relative velocities of interfaces and layers. As settling continues layers of lower concentration follow layers of higher concentration. A descending layer or interface will only appear if it falls faster than and 'gets away from' the more dilute layers descending from above them. In simple terms, if an interface between zone B and a suspension of concentration greater than $C_B$ but less that $C_S$ descends faster than the interface between zones B and S then a zone of variable concentration will form. Examination of the particle flux plot enables us to determine which type of settling is occurring. Referring to Figure 2.8, a tangent to the curve is drawn through the point ($C = C_S$, $U_{ps} = 0$). The concentration at the point of tangent is $C_{B_2}$. The concentration at the point of intersection of the projected tangent with the curve is $C_{B_1}$. Type 1 settling occurs when the initial suspension concentration is less than $C_{B_1}$. Type 2 settling occurs when the initial suspension concentration lies between $C_{B_1}$ and $C_{B_2}$.

### 2.2.4 Relationship Between the Height–Time Curve and the Flux Plot

Following the AB interface in the simple batch settling test gives rise to the height–time curve shown in Figure 2.9. In fact, there will be a family of such curves for different initial concentrations. The following analysis permits the derivation of the particle flux plot from the height–time curve.

Referring to Figure 2.9, at time $t$ the interface between clear liquid and

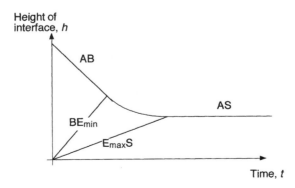

**Figure 2.7**   Change in positions of interface AB, $BE_{min}$, $E_{max}S$ and AS with time in type 2 batch settling (e.g. AB is the interface between zone A and zone B. $BE_{min}$ is the interface between zone B and the lowest suspension concentration in the variable zone E; see Figure 2.6)

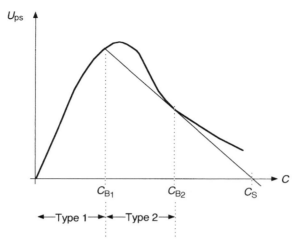

**Figure 2.8** Determining if settling will be type 1 or type 2. A line through $C_S$ tangent to the flux curve gives $C_{B_1}$ and $C_{B_2}$. Type 2 settling occurs when initial suspension concentration is between $C_{B_1}$ and $C_{B_2}$

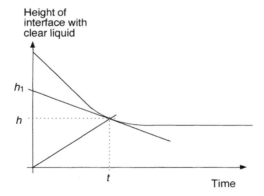

**Figure 2.9** Analysis of batch settling test

suspension of concentration, $C$ is at a height $h$ from the base of the vessel and velocity of the interface is the slope of the curve at this time:

$$\text{velocity of interface} = \frac{\mathrm{d}h}{\mathrm{d}t} = \frac{h_1 - h}{t} \tag{2.32}$$

this is also equal to $U_p$, the velocity of the particles at the interface relative to the vessel wall. Hence,

$$U_p = \frac{h_1 - h}{t} \tag{2.33}$$

Now consider planes or waves of higher concentration which rise from the base of the vessel. At time, $t$, a plane of concentration, $C$ has risen a distance $h$ from the base. Thus the velocity at which a plane of concentration, $C$, rises from the base is $h/t$. This plane or wave of concentration passes up through the suspension. The velocity of the particles relative to the plane is therefore:

$$\text{velocity of particles relative to plane} = U_p + \frac{h}{t} \tag{2.34}$$

As the particles pass through the plane they have a concentration, $C$ (refer to Figure 2.10). Therefore, the volume of particles which have passed through this plane in time $t$ is

$$= \text{area} \times \text{velocity of particles} \times \text{concentration} \times \text{time}$$

$$= A\left(U_p + \frac{h}{t}\right)Ct \tag{2.35}$$

But, at time $t$ this plane is interfacing with the clear liquid, and so at this time all the particles in the test have passed through the plane.

$$\text{The total volume of all the particles in the test} = C_B h_0 A \tag{2.36}$$

(where, $h_0$ is the initial suspension height)
   Therefore,

$$C_B h_0 A = A\left(U_p + \frac{h}{t}\right)Ct \tag{2.37}$$

hence, substituting for $U_p$ from Equation (2.33), we have

$$C = \frac{C_B h_0}{h_1} \tag{2.38}$$

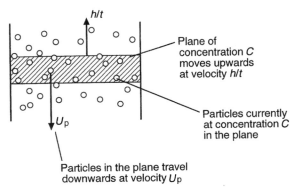

**Figure 2.10**   Analysis of batch settling; relative velocities of a plane of concentration $C$ and the particles in the plane.

## 2.3 CONTINUOUS SETTLING

### 2.3.1 Settling of a Suspension in a Flowing Fluid

We will now look at the effects of imposing a net fluid flow on to the particle settling process with a view to eventually producing a design procedure for a thickener. This analysis follows the method suggested by Fryer and Uhlherr (1980).

Firstly, we will consider a settling suspension flowing downwards in a vessel. A suspension of solids concentration, $(1 - \varepsilon_F)$ or $C_F$ is fed continuously into the top of a vessel of cross sectional area, $A$ at a volume flow rate, $Q$ (Figure 2.11). The suspension is drawn off from the base of the vessel at the same rate. At a given axial position, $X$, in the vessel let the local solids concentration be $(1 - \varepsilon)$ or $C$ and the volumetric fluxes of the solids and the fluid be $U_{ps}$ and $U_{fs}$ respectively. Then assuming incompressible fluid and solids, continuity gives

$$Q = (U_{ps} + U_{fs})A \tag{2.39}$$

At position $X$, the relative velocity between fluid and particles, $U_{rel}$ is given by

$$U_{rel} = \frac{U_{ps}}{1 - \varepsilon} - \frac{U_{fs}}{\varepsilon} \tag{2.40}$$

Our analysis of batch settling gave us the following expression for this relative velocity:

$$U_{rel} = U_T \varepsilon f(\varepsilon) \tag{2.7}$$

and so, combining Equations (2.39), (2.40) and (2.7) we have

$$U_{ps} = \frac{Q(1 - \varepsilon)}{A} + U_T \varepsilon^2 (1 - \varepsilon) f(\varepsilon) \tag{2.41}$$

or

total solids flux = flux due to bulk flow + flux due to settling

We can use this expression to convert our batch flux plot into a continuous total downward flux plot. Referring to Figure 2.12, we plot a line of slope $Q/A$

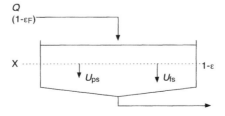

**Figure 2.11**  Continuous settling; downflow only

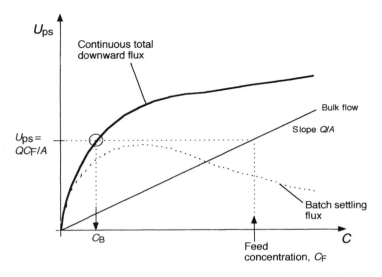

**Figure 2.12**   Total flux plot for settling in downward flow

through the origin to represent the bulk flow flux and then add this to the batch flux plot to give the continuous total downward flux plot. Now, in order to graphically determine the solids concentration at level X in the vessel we apply the mass balance between feed and the point X. Reading up from the feed concentration $C_F$ to the bulk flow line gives the value of the volumetric particle flux fed to the vessel, $QC_F/A$. By continuity this must also be the total flux at level X or any level in the vessel. Hence, reading across from the flux of $QC_F/A$ to the continuous total flux curve, we may read off the particle concentration in the vessel during downward flow, which we will call $C_B$ (The subscript B will eventually refer to the 'bottom' section of the continuous thickener). In downward flow the value of $C_B$ will always be lower than the feed concentration $C_F$, since the solids velocity is greater in downward flow than in the feed (concentration × velocity = flux).

A similar analysis applied to upward flow of a particle suspension in a vessel gives: total downward particle flux,

$$U_{ps} = U_T \varepsilon^2 f(\varepsilon) - \frac{Q(1-\varepsilon)}{A} \qquad (2.42)$$

or

total solids flux = flux due to settling − flux due to bulk flow

Hence, for upward flow, we obtain the continuous total flux plot by sub-tracting the straight line representing the flux due to bulk flow from the batch flux curve (Figure 2.13). Applying the material balance as we did for down-ward flow, we are able to graphically determine the particle concentration in the vessel during upward flow of fluid, $C_T$ (the T refers to the 'top' section of

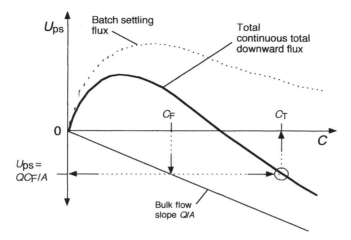

**Figure 2.13**   Total flux plot for settling in upward flow

the continuous thickener). It will be seen from Figure 2.13 that the value of particle concentration for the upward-flowing suspension, $C_T$, is always greater than the feed concentration, $C_F$. This is because the particle velocity during upward flow is always less than that in the feed.

## 2.3.2   A Real Thickener (with Upflow and Downflow Sections)

Consider now a real thickener shown schematically in Figure 2.14. The feed suspension of concentration $C_F$ is fed into the vessel at some point intermediate between the top and bottom of the vessel at a volume flow rate, $F$. An 'underflow' is drawn off at the base of the vessel at a volume flow rate, $L$, and concentration $C_L$. A suspension of concentration $C_V$ overflows at a volume flow rate $V$ at the top of the vessel (this flow is called the 'overflow'). Let the mean particle concentrations in the bottom (downflow) and top (upflow) sections be $C_B$ and $C_T$ respectively. The total and particle material balances over the thickener are:

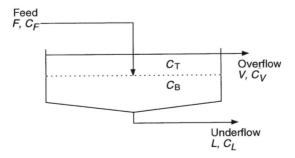

**Figure 2.14**   A real thickener, combining upflow and downflow ($F$, $L$ and $V$ are volume flows; $C_F$, $C_L$ and $C_V$ are concentrations)

Total:

$$F = V + L \tag{2.43}$$

Particle:

$$FC_F = VC_V + LC_L \tag{2.44}$$

These material balances link the total continuous flux plots for the upflow and downflow sections in the thickener.

### 2.3.3 Critically Loaded Thickener

Figure 2.15 shows flux plots for a 'critically loaded' thickener. The line of slope $F/A$ represents the relationship between feed concentration and feed flux for a

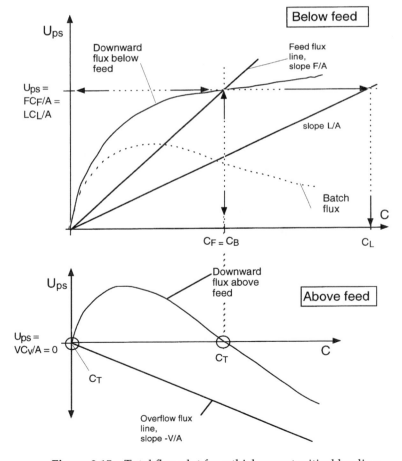

**Figure 2.15**   Total flux plot for a thickener at critical loading

volumetric feed rate, $F$. The material balance Equations, (2.43) and (2.44), determine that this line intersects the curve for the total flux in the downflow section when the total flux in the upflow section is zero. Under critical loading conditions the feed concentration is just equal to the critical value giving rise to a feed flux equal to the total continuous flux that the downflow section can deliver at that concentration. Thus the combined effect of bulk flow and settling in the downflow section provides a flux equal to that of the feed. Under these conditions, since all particles fed to the thickener can be dealt with by the downflow section, the upflow flux is zero. The material balance then dictates that the concentration in the downflow section, $C_B$, is equal to $C_F$ and the underflow concentration, $C_L$ is $FC_F/L$. The material balance may be performed graphically and is shown on Figure 2.15. From the feed flux line, the feed flux at a feed concentration, $C_F$ is $U_{ps} = FC_F/A$. At this flux the concentration in the downflow section is $C_B = C_F$. The downflow flux is exactly equal to the feed flux and so the flux in the upflow section is zero. In

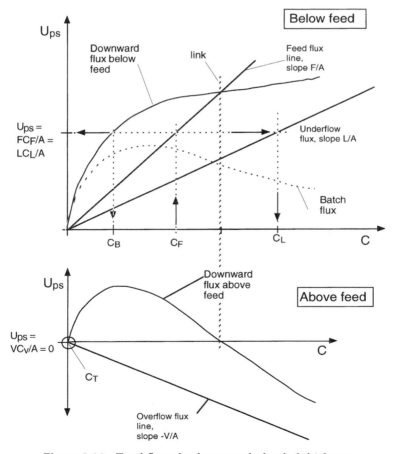

**Figure 2.16**   Total flux plot for an underloaded thickener

the underflow, where there is no sedimentation, the underflow flux, $LC_L/A$, is equal to the downflow flux. At this flux the underflow concentration, $C_L$ is determined from the underflow line.

Figure 2.15 indicates that under critical conditions there are two possible solu-tions for the concentration in the upflow section, $C_T$. One solution, the obvious one, is $C_T = 0$; the other is $C_T = C_B$. In this second situation a fluidized bed of particles at concentration $C_B$ with a distinct surface is observed in the upflow section.

### 2.3.4 Underloaded Thickener

When the feed concentration $C_F$ is less than the critical concentration the thickener is said to be underloaded. This situation is depicted in Figure 2.16. Here the feed flux, $FC_F/A$, is less than the maximum flux due to bulk flow and settling which can be provided by the downflow section. The flux in the upflow section is again zero ($C_T = C_V = 0$; $VC_V/A = 0$). The graphical mass balance shown in Figure 2.16 enables $C_B$ and $C_L$ to be determined (feed flux = downflow section flux = underflow flux).

### 2.3.5 Overloaded Thickener

When the feed concentration $C_F$ is greater than the critical concentration, the thickener is said to be overloaded. This situation is depicted in Figure 2.17. Here the feed flux, $FC_F/A$, is greater than the maximum flux due to bulk flow and settling provided by the downflow section. The excess flux must pass through the upflow section and out through the overflow. The graphical material balance is depicted in Figure 2.17. At the feed concentration $C_F$, the difference between the feed flux and the total flux in the downflow section gives the excess flux which must pass through the upflow section. This flux applied to the upflow section graph gives the value of the concentration in the upflow section, $C_T$, and the overflow concentration, $C_V$ (upflow section flux = overflow flux).

### 2.3.6 Alternative Form of Total Flux Plot

A common form of continuous flux plot is that exhibiting a minimum total flux shown under critical conditions in Figure 2.18. With this alternative flux plot the critical loading condition occurs when the feed concentration gives rise to a flux equal to this minimum in the total flux curve. The downflow section cannot operate in a stable manner above this flux. Under critical conditions the upflow flux is again zero and the graphical material balance depicted in Figure 2.18 gives the values of $C_B$ and $C_L$. It will be noted that under these conditions there are two possible values of $C_B$; these may coexist in the downflow section with a discontinuity between them at any position between the feed level and the underflow.

Figure 2.19 shows this alternative flux plot in an overloaded situation. For

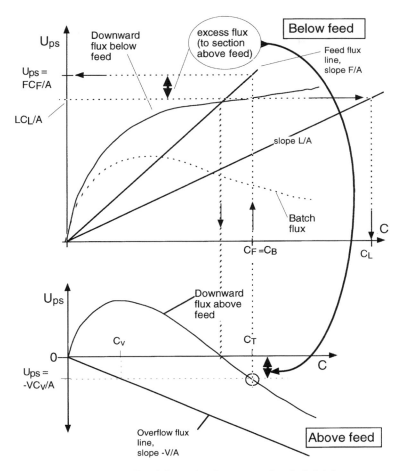

**Figure 2.17** Total flux plot for an overloaded thickener

the graphical solution in this case, the excess flux must be read from the flux axis of the downflow section plot and applied to the upflow section plot in order to determine the value of $C_T$ and $C_V$. Note that in this case although there are theoretically two possible values of $C_B$, in practice only the higher value can stably coexist with the higher concentration region, $C_T$, above it.

## 2.4 WORKED EXAMPLES

### WORKED EXAMPLE 2.1

A height–time curve for the sedimentation of a suspension, of initial suspension concentration 0.1, in vertical cylindrical vessel is shown in Figure 2W1.1. Determine:

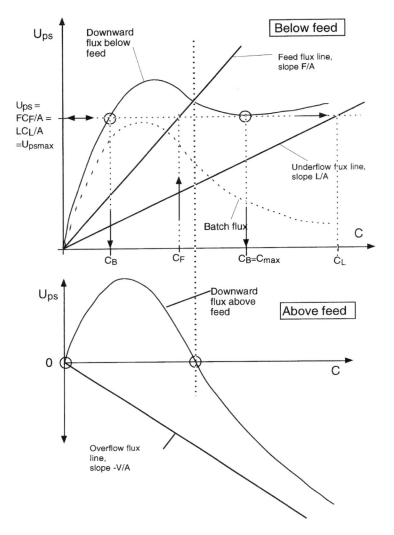

**Figure 2.18** Alternative total flux plot shape; thickener at critical loading

(a) the velocity of the interface between clear liquid and suspension of concentration 0.1.

(b) the velocity of the interface between clear liquid and a suspension of concentration 0.175.

(c) the velocity at which a layer of concentration 0.175 propagates upwards from the base of the vessel.

(d) the final sediment concentration.

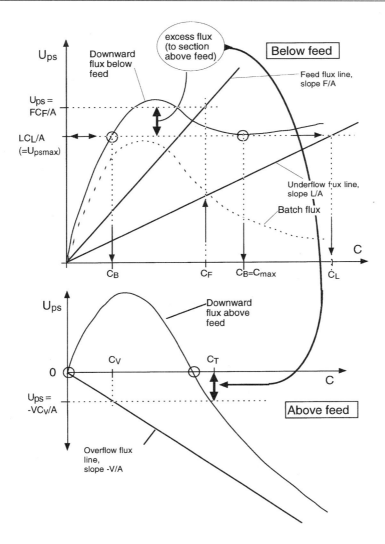

**Figure 2.19**  Alternative total flux plot shape; overloaded thickener

## Solution

(a)  Since the initial suspension concentration is 0.1, the velocity required in this question is the velocity of the AB interface. This is given by the slope of the straight portion of the height–time curve.

$$\text{Slope} = \frac{20 - 40}{15 - 0} = 1.333 \text{ cm/s}$$

(b)  We must first find the point on the curve corresponding to the point at which a suspension of concentration 0.175 interfaces with the clear suspension. From Equation (2.38), with $C = 0.175$, $C_B = 0.1$ and $h_0 = 40$ cm, we find:

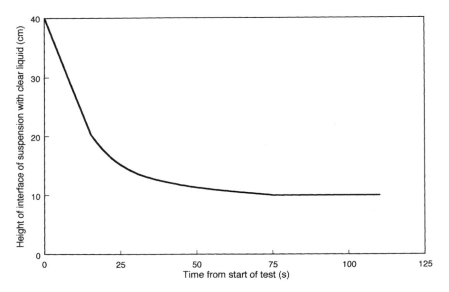

**Figure 2.W1.1**   Batch settling test; height–time curve

$$h_1 = \frac{0.1 \times 40}{0.175} = 22.85 \text{ cm}$$

A line drawn through the point $t = 0$, $h = h_1$ tangent to the curve locates the point on the curve corresponding to the time at which a suspension of concentration 0.175 interfaces with the clear suspension (Figure 2.W1.2). The

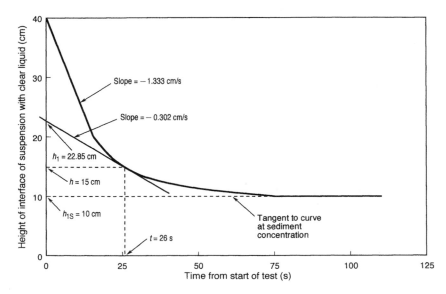

**Figure 2.W1.2**   Batch settling test. Solution to Worked Example 2.1

coordinates of this point are $t = 26$ s, $h = 15$ cm. The velocity of this interface is the slope of the curve at this point:

$$\text{slope of curve at 26 s, 15 cm} = \frac{15 - 22.85}{26 - 0} = -0.302 \text{ cm/s}$$

downward velocity of interface $= 0.30$ cm/s

(c)   From the consideration above, after 26 s the layer of concentration 0.175 has just reached the clear liquid interface and has travelled a distance of 15 cm from the base of the vessel in this time.

Therefore, upward propagation velocity of this layer $= \dfrac{h}{t} = \dfrac{15}{26} = 0.577$ cm/s

(d)   To find the concentration of the final sediment we again use Equation (2.38). The value of $h_1$ corresponding to the final sediment ($h_{1S}$) is found by drawing a tangent to the part of the curve corresponding to the final sediment and projecting it to the $h$ axis.
   In this case $h_{1S} = 10$ cm and so from Equation (2.38),

$$\text{final sediment concentration, } C = \frac{C_0 h_0}{h_{1S}} = \frac{0.1 \times 4.0}{10} = 0.4$$

## WORKED EXAMPLE 2.2

A suspension in water of uniformly sized sphere (diameter 150 μm, density 1140 kg/m$^3$) has a solids concentration of 25% by volume. The suspension settles to a bed of solids concentration of 62% by volume. Calculate:

(a)   the rate at which the water/suspension interface settles;

(b)   the rate at which the sediment/suspension interface rises. (assume water properties: density, 1000 kg/m$^3$; viscosity, 0.001 Pa s)

## Solution

(a) Solids concentration of initial suspension, $C_B = 0.25$
Equation (2.28) allows us to calculate the velocity of interfaces between suspensions of different concentrations:
The velocity of the interface between initial suspension (B) and clear liquid (A) is therefore:

$$U_{\text{int,AB}} = \frac{U_{pA}C_A - U_{pB}C_B}{C_A - C_B}$$

Since $C_A = 0$, the equation reduces to

$$U_{\text{int,AB}} = U_{\text{pB}}$$

$U_{\text{pB}}$ is the hindered settling velocity of particles relative to the vessel wall in batch settling and is given by Equation (2.24):

$$U_{\text{p}} = U_{\text{T}} \varepsilon^n$$

Assuming Stokes law applies, then $n = 4.65$ and the single particle terminal velocity is given by Equation (1.13) (see Chapter 1):

$$U_{\text{T}} = \frac{x^2(\rho_{\text{p}} - \rho_{\text{f}})g}{18\mu}$$

$$U_{\text{T}} = \frac{9.81 \times (150 \times \times 10^{-6})^2 \times (1140 - 1000)}{18 \times 0.001}$$

$$= 1.717 \times 10^{-3} \text{ m/s}$$

To check that the assumption of Stokes' law is valid, we calculate the single particle Reynolds number:

$$Re_{\text{p}} = \frac{(150 \times 10^{-3}) \times 1.717 \times 10^{-3} \times 1000}{0.001}$$

$= 0.258$, which is less than the limiting value for Stokes' law (0.3) and so the assumption is valid.

The voidage of the initial suspension, $\varepsilon_{\text{B}} = 1 - C_{\text{B}} = 0.75$

hence,     $U_{\text{pB}} = 1.717 \times 10^{-3} \times 0.75^{4.65}$

$$= 0.45 \times 10^{-3} \text{ m/s}$$

Hence, the velocity of the interface between the initial suspension and the clear liquid is 0.45 mm/s. The fact that the velocity is positive indicates that the interface is moving downwards.

(b) Here again we apply Equation (2.28) to calculate the velocity of interfaces between suspensions of different concentrations:

The velocity of the interface between initial suspension (B) and sediment (S) is therefore

$$U_{\text{int,BS}} = \frac{U_{\text{pB}} C_{\text{B}} - U_{\text{pS}} C_{\text{S}}}{C_{\text{B}} - C_{\text{S}}}$$

With $C_{\text{B}} = 0.25$ and $C_{\text{S}} = 0.62$ and since the velocity of the sediment, $U_{\text{pS}}$ is zero, we have:

$$U_{\text{int,BS}} = \frac{U_{\text{pB}} 0.25 - 0}{0.25 - 0.62} = -0.675 U_{\text{pB}}$$

And from part (a), we know that $U_{\text{pB}} = 0.45$ mm/s, and so $U_{\text{int,BD}} = -0.304$ mm/s. The negative sign signifies that the interface is moving upwards. So, the interface between initial suspension and sediment is moving upwards at a velocity of 0.304 mm/s.

## WORKED EXAMPLE 2.3

For the batch flux plot shown in Figure 2.W3.1, the sediment has a solids concentration of 0.4 volume fraction of solids.

(a) Determine the range of initial suspension concentrations over which a zone of variable concentration is formed under batch settling conditions.

(b) Calculate and plot the concentration profile after 50 minutes in a batch settling test of a suspension with an initial concentration 0.1 volume fraction of solids, and initial suspension height of 100 cm.

(c) At what time will the settling test be complete?

### Solution

(a) Determine the range of initial suspension concentrations by drawing a line through the point $C = C_S = 0.4$, $U_{ps} = 0$ tangent to the batch flux curve. This is shown as line $XC_S$ in Figure 2.W3.2. The range of initial suspension concentrations for which a zone of variable concentration is formed in batch settling (type 2 settling) is defined by $C_{B_{min}}$ and $C_{B_{max}}$. $C_{B_{min}}$ is the value of $C$ at which the line $XC_S$ intersects the settling curve and $C_{B_{max}}$ is the value of $C$ at the tangent. From Figure 2W3.2, we see that $C_{B_{min}} = 0.036$ and $C_{B_{max}} = 0.21$.

(b) To calculate the concentration profile we must first determine the velocities of the interfaces between the zones A, B, E and S and hence find their positions after 50 minutes.

The line AB in Figure 2.W3.2 joins the point representing A the clear liquid (0, 0) and the point B representing the initial suspension (0.1, $U_{ps}$). The slope of line AB

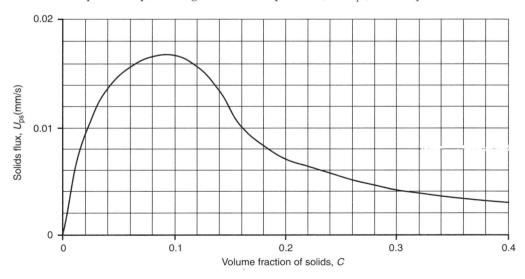

**Figure 2.W3.1** Batch flux plot for Worked Example 2.3

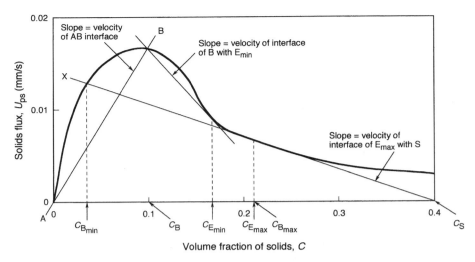

**Figure 2.W3.2**   Graphical solution to batch settling problem in Worked Example 2.3

is equal to the velocity of the interface between zones A and B. From Figure 2.W3.2, $U_{\text{int,AB}} = +0.166$ mm/s or $+1.00$ cm/min

The slope of the line from point B tangent to the curve is equal to the velocity of the interface between the initial suspension B and the minimum value of the variable concentration zone $C_{\text{Emin}}$.

From Figure 2.W3.2,

$$U_{\text{int,BE}_{\text{min}}} = -0.111 \text{ mm/s or } -0.64 \text{ cm/min.}$$

The slope of the line tangent to the curve and passing through the point representing the sediment (point $C = C_S = 0.4$, $U_{\text{ps}} = 0$) is equal to the velocity of the interface between the maximum value of the variable concentration zone $C_{\text{Emax}}$ and the sediment.

From Figure 2.W3.2,

$$U_{\text{int,E}_{\text{max}}S} = -0.0355 \text{ mm/s or } -0.213 \text{ cm/min.}$$

Therefore, after 50 minutes the distances travelled by the interfaces will be

| | |
|---|---|
| AB interface | 50.0 cm ($1.00 \times 50$) downwards |
| BE$_{\text{min}}$ interface | 33.2 cm upwards |
| E$_{\text{max}}$S interface | 10.6 cm upwards |

Therefore, the positions of the interfaces (distance from the base of the test vessel) after 50 minutes will be

| | |
|---|---|
| AB interface | 50.0 cm |
| BE$_{\text{min}}$ interface | 33.2 cm |
| E$_{\text{max}}$S interface | 10.6 cm |

From Figure 2.W3.2 we determine the minimum and maximum values of suspension concentration in the variable zone

$$C_{E_{min}} = 0.16$$

$$C_{E_{max}} = 0.21$$

Using this information we can plot the concentration profile in the test vessel 50 min after the start of the test. A sketch of the profile is shown in Figure 2.W3.3. The shape of the concentration profile within the variable concentration zone may be determined by the following method. Recalling that the slope of the batch flux plot (Figure 2.W3.1) at a value of suspension concentration $C$ is the velocity of a layer of suspension of that concentration, we find the slope at two or more values of concentration and then determine the positions of these layers after 50 min:

- Slope of batch flux plot at $C = 0.18$ is 0.44 cm/min upwards.
  Hence, position of a layer of concentration 0.18 after 50 min is 22.0 cm from the base.

- Slope of batch flux plot at $C = 0.20$, is 0.27 cm/min upwards.
  Hence, position of a layer of concentration 0.20 after 50 min is 13.3 cm from the base.

These two points are plotted on the concentration profile in order to determine the shape of the profile within the zone of variable concentration.

Figure 2.W3.4 is a sketched plot of the height–time curve for this test constructed from the information above. The shape of the curved portion of the curve can again be determined by plotting the positions of two or more layers of suspension of different concentration. The initial suspension concentration zone

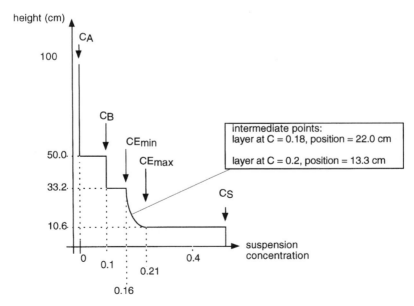

**Figure 2.W3.3** Sketch of concentration profile in batch settling test vessel after 50 minutes. Worked Example 2.3

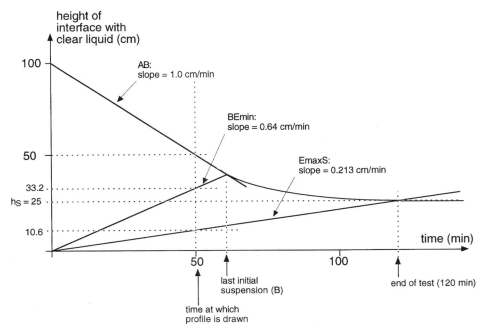

**Figure 2.W3.4**  Sketch of height–time curve for the batch settling test in Worked Example 2.3

(B) ends when the AB line intersects the $BE_{min}$ line, both of which are plotted from a knowledge of their slopes.

The time for the end of the test is found in the following way. The end of the test is when the position of the $E_{max}S$ interface coincides with the height of the final sediment. The height of the final sediment may be found using Equation (2.38) (see part (d) of Worked Example 2.1):

$$C_S h_S = C_B h_0$$

where $h_S$ is the height of the final sediment and $h_0$ is the initial height of the suspension (at the start of the test). With $C_S = 0.4$, $C_B = 0.1$ and $h_0 = 100$ cm, we find that $h_S = 25$ cm. Plotting $h_s$ on Figure 2.W3.4, we find that the $E_{max}S$ line intersects the final sediment line at about 120 min and so the test ends at this time.

## WORKED EXAMPLE 2.4

Using the flux plot shown in Figure 2.W4.1, (a) graphically determine the limiting feed concentration for a thickener of area 100 m$^2$ handling a feed rate of 0.019 m$^3$/s and an underflow rate of 0.01 m$^3$/s. Under these conditions what will be the underflow concentration and the overflow concentration?

(b) Under the same flow conditions as above, the feed concentration is increased to 0.2. Estimate the solids concentration in the overflow, in the underflow, in the upflow section and in the downflow section of the thickener.

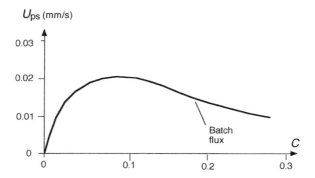

**Figure 2.W4.1** Batch flux plot for Worked Example 2.4

## Solution

(a) Feed rate, $F = 0.019$ m$^3$/s
Underflow rate, $L = 0.01$ m$^3$/s
   Material balance gives, overflow rate, $V = F - L = 0.009$ m$^3$/s
Expressing these flows as fluxes based on the thickener area ($A = 100$ m$^2$):

$$\frac{F}{A} = 0.19 \text{ mm/s}$$

$$\frac{L}{A} = 0.10 \text{ mm/s}$$

$$\frac{V}{A} = 0.09 \text{ mm/s}$$

The relationships between bulk flux and suspension concentration are then:

$$\text{Feed flux} = C_F \left( \frac{F}{A} \right)$$

$$\text{Flux in underflow} = C_L \left( \frac{L}{A} \right)$$

$$\text{Flux in overflow} = C_V \left( \frac{V}{A} \right)$$

   Lines of slope $F/A$, $L/A$ and $-V/A$ drawn on the flux plot represent the fluxes in the feed, underflow and overflow respectively (Figure 2.W4.2). The total flux plot for the section below the feed point is found by adding batch flux plot to the underflow flux line. The total flux plot for the section above the feed point is found by adding the batch flux plot to the overflow flux line (which is negative since it is an upward flux). These plots are shown in Figure 2.W4.2.
   The critical feed concentration is found where the feed flux line intersects the plot of total flux in the section below the feed (Figure 2.W4.2). This gives a critical feed flux of 0.0335 mm/s. The downflow section below the feed point is unable to take a flux greater than this. The corresponding feed concentration is $C_{F_{crit}} = 0.174$.
   The concentration in the downflow section, $C_B$ is also 0.174.

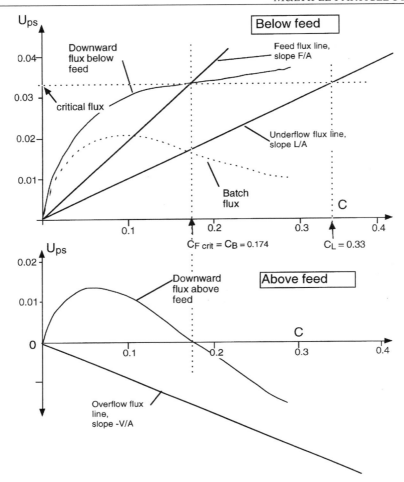

**Figure 2.W4.2**   Total flux plot: solution to part (a) of Worked Example 2.4

The corresponding concentration in the underflow is found where the critical flux line intersects the underflow flux line. This gives $C_L = 0.33$.

(b) Referring now to Figure 2.W4.3, if the feed flux is increased to 0.2, we see that the corresponding feed flux is 0.038 mm/s. At this feed concentration the down-flow section is only able to take a flux of 0.034 mm/s and gives an underflow concentration, $C_L = 0.34$. The excess flux of 0.004 mm/s passes into the upflow section. This flux in the upflow section gives a concentration, $C_T = 0.2$ and a corresponding concentration, $C_V = 0.044$ in the overflow.

## EXERCISES

**2.1** A suspension in water of uniformly sized spheres of diameter 100 μm and density 1200 kg/m$^3$ has a solids volume fraction of 0.2. The suspension settles to a

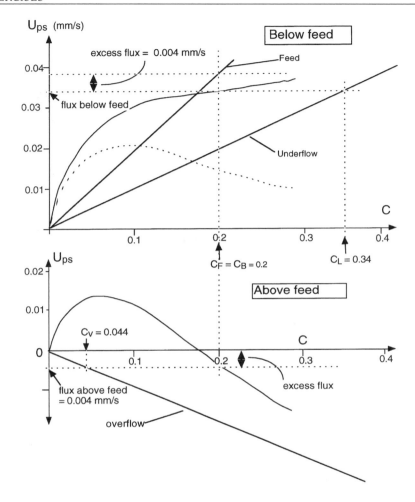

**Figure 2.W4.3** Solution to part (b) of Worked Example 2.4

bed of solids volume fraction 0.5. (For water, density is 1000 kg/m$^3$ and viscosity is 0.001 Pa s)

The single particle terminal velocity of the spheres in water may be taken as 1.1 mm/s.

Calculate

(a) the velocity at which the clear water/suspension interface settles;

(b) the velocity at which the sediment/suspension interface rises.

(Answers: (a) 0.39 mm/s; (b) 0.26 mm/s)

**2.2** A height–time curve for the sedimentation of a suspension in a vertical cylindrical vessel is shown in Figure 2.E2.1. The initial solids concentration of the suspension is 150 kg/m$^3$.

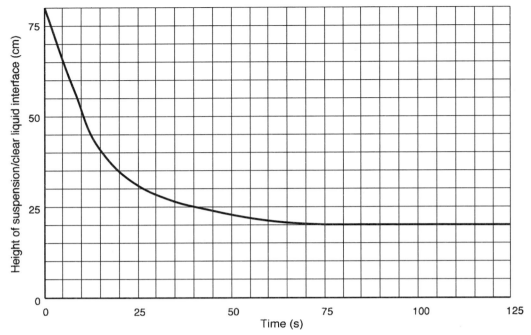

**Figure 2.E2.1**   Batch settling test results. Height versus time curve for use in Questions 2.2 and 2.4

Determine:

(a)   The velocity of the interface between clear liquid and suspension of concentration 150 kg/m$^3$.

(b)   The time from the start of the test at which the suspension of concentration 240 kg/m$^3$ is in contact with the clear liquid.

(c)   The velocity of the interface between the clear liquid and suspension of concentration 240 kg/m$^3$.

(d)   The velocity at which a layer of concentration 240 kg/m$^3$ propagates upwards from the base of the vessel.

(e)   The concentration of the final sediment.

(Answers: (a) 2.91 cm/s; (b) 22 s; (c) 0.77 cm/s downwards (d) 1.50 cm/s upwards; (e) 600 kg/m$^3$).

**2.3**  A suspension in water of uniformly sized spheres of diameter 90 μm and density 1100 kg/m$^3$ has a solids volume fraction of 0.2. The suspension settles to a bed of solids volume fraction 0.5. (For water: density is 1000 kg/m$^3$ and viscosity is 0.001 Pa s).

The single particle terminal velocity of the spheres in water may be taken as 0.44 mm/s.

Calculate:

(a)   The velocity at which the clear water/suspension interface settles;

(b)   The velocity at which the sediment/suspension interface rises.

(Answers: (a) 0.156 mm/s; (b) 0.104 mm/s)


**2.4** A height–time curve for the sedimentation of a suspension in a vertical cylindrical vessel is shown in Figure 2.E2.1. The initial solids concentration of the suspension is 200 kg/m$^3$.
   Determine:

(a)   the velocity of the interface between clear liquid and suspension of concentration 200 kg/m$^3$;

(b)   the time from the start of the test at which the suspension of concentration 400 kg/m$^3$ is in contact with the clear liquid;

(c)   the velocity of the interface between the clear liquid and suspension of concentration 400 kg/m$^3$;

(d)   the velocity at which a layer of concentration 400 kg/m$^3$ propagates upwards from the base of the vessel;

(e)   the concentration of the final sediment.

(Answers: (a) 2.67 cm/s downwards; (b) 35 s; (c) 0.386 cm/s downwards; (d) 0.846 cm/s downwards (e) 800 kg/m$^3$.)


**2.5**
(a)   Spherical particles of uniform diameter 40 µm and particle density 2000 kg/m$^3$ form a suspension of solids volume fraction 0.32 in a liquid of density 880 kg/m$^3$ and viscosity 0.0008 Pa s. Assuming Stokes' law applies, calculate (i) the sedimentation velocity and (ii) the sedimentation volumetric flux for this suspension.

(b)   A height–time curve for the sedimentation of a suspension in a cylindrical vessel is shown in Figure 2.E5.1. The initial concentration of the suspension for this test is 0.12 m$^3$/m$^3$. Calculate:

   (i) the velocity of the interface between clear liquid and a suspension of concentration, 0.12 m$^3$/m$^3$.

   (ii) the velocity of the interface between clear liquid and a suspension of concentration 0.2 m$^3$/m$^3$.

   (iii) the velocity at which a layer of concentration, 0.2 m$^3$/m$^3$ propagates upwards from the base of the vessel.

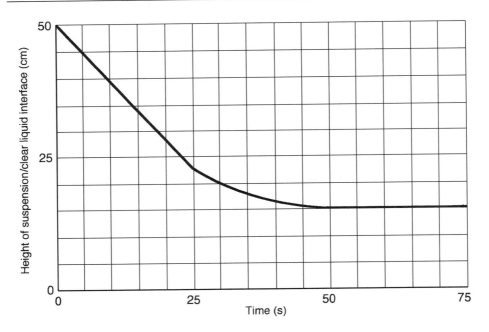

**Figure 2.E5.1**  Batch settling test results. Height versus time curve for use in Question 2.5

(iv)  The concentration of the final sediment.

(v)  The velocity at which the sediment propagates upwards from the base.

(Answers:  (a)(i)  0.203 mm/s;  (ii)  0.065 mm/s;  (b)(i)  1.11 cm/s  downwards;  (ii) 0.345 cm/s downwards, (iii) 0.514 cm/s upwards, (iv) 0.4; (v) 0.30 cm/s upwards).

**2.6**  A height–time curve for the sedimentation of a suspension in a vertical cylindrical vessel is shown in Figure 2.E6.1. The initial solids concentration of the suspension is $100 \, kg/m^3$.
    Determine:

(a)  the velocity of the interface between clear liquid and suspension of concentration $100 \, kg/m^3$;

(b)  the time from the start of the test at which the suspension of concentration $200 \, kg/m^3$ is in contact with the clear liquid;

(c)  the velocity of the interface between the clear liquid and suspension of concentration $200 \, kg/m^3$;

(d)  the velocity at which a layer of concentration $200 \, kg/m^3$ propagates upwards from the base of the vessel;

(e)  the concentration of the final sediment.

(Answers: (a) 0.667 cm/s downwards; (b) 140 s; (c) 0.0976 cm/s downwards; (d) 0.189 cm/s upwards; (e) $400 \, kg/m^3$.)

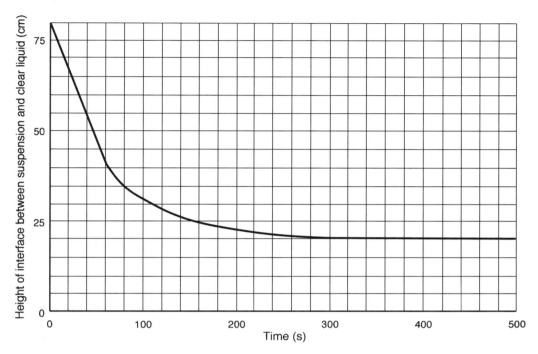

**Figure 2.E6.1** Batch settling test results. Height versus time curve for use in Questions 2.6 and 2.8

**2.7** A suspension in water of uniformly sized spheres of diameter 80 μm and density 1300 kg/m³ has a solids volume fraction of 0.10. The suspension settles to a bed of solids volume fraction 0.4. (For water, density is 1000 kg/m³ and viscosity is 0.001 Pa s)

The single particle terminal velocity of the spheres under these conditions is 1.0 mm/s.

Calculate:

(a)  the velocity at which the clear water/suspension interface settles;

(b)  the velocity at which the sediment/suspension interface rises.

(Answers: (a) 0.613 mm/s; (b) 0.204 mm/s.)

**2.8** A height–time curve for the sedimentation of a suspension in a vertical cylindrical vessel is shown in Figure 2.E6.1. The initial solids concentration of the suspension is 125 kg/m³.

Determine:

(a)  The velocity of the interface between clear liquid and suspension of concentration 125 kg/m³.

(b)  The time from the start of the test at which the suspension of concentration 200 kg/m³ is in contact with the clear liquid.

(c)    The velocity of the interface between the clear liquid and suspension of concentration 200 kg/m$^3$.

(d)    The velocity at which a layer of concentration 200 kg/m$^3$ propagates upwards from the base of the vessel.

(e)    The concentration of the final sediment.

(Answers: (a) 0.667 cm/s downwards; (b) 80 s; (c) 0.192 cm/s downwards (d) 0.438 cm/s upwards (e) 500 kg/m$^3$.)

**2.9**  Use the batch flux plot in Figure 2.E9.1 to answer the following questions (Note that the sediment concentration is 0.44 volume fraction).

(a)    Determine the range of initial suspension concentration over which a variable concentration zone is formed under batch settling conditions.

(b)    For a batch settling test using a suspension with an initial concentration 0.18 volume fraction and initial height 50 cm, determine the settling velocity of the interface between clear liquid and suspension of concentration 0.18 volume fraction.

(c)    Determine the position of this interface 20 min after the start of this test.

(d)    Produce a sketch showing the concentration zones in the settling test 20 minutes after the start of this test.

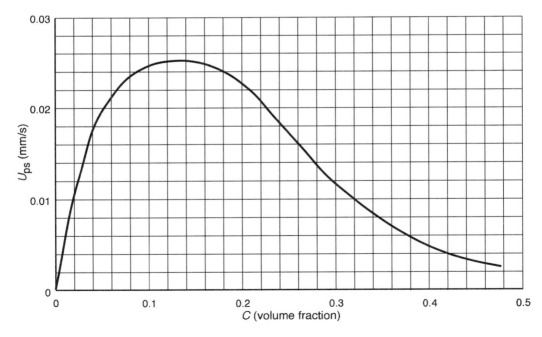

**Figure 2.E9.1**  Batch flux plot for use in Question 2.9

(Answers: (a) 0.135 to 0.318; (b) 0.80 cm/min; (c) 34 cm from base; (d) BE interface is 12.5 cm from base.)

**2.10** Consider the batch flux plot shown in Worked Example 3 (Figure 2.W3.1). Given that the final sediment concentration is 0.36 volume fraction,

(a)   determine the range of initial suspension concentration over which a variable concentration zone is formed under batch settling conditions;

(b)   calculate and sketch the concentration profile after 40 minutes of the batch settling test with an initial suspension concentration of 0.08 and an initial height of 100 cm;

(c)   estimate the height of the final sediment and the time at which the test is complete.

(Answers: (a) 0.045 to 0.20; (c) 22.2 cm; 83 minutes)

**2.11** The batch and continuous flux plots supplied in Figure 2.E11.1 are for a thickener of area 200 m$^2$ handling a feed rate of 0.04 m$^3$/s and an underflow rate of 0.025 m$^3$/s.

(a)   Using these plots, graphically determine the critical or limiting feed concentration for this thickener.

(b)   Given that if the feed concentration is 0.18 m$^3$/m$^3$, determine the solids concentrations in the overflow, underflow, in the regions above and below the feed well.

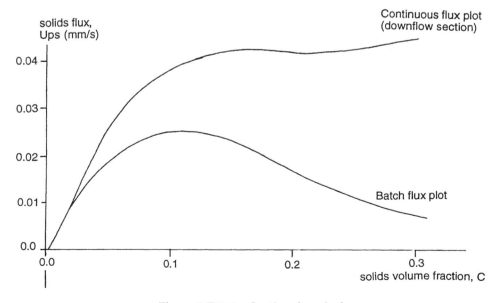

**Figure 2.E11.1**   *Continued overleaf*

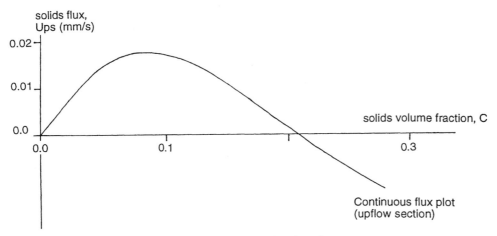

**Figure 2.E11.1** *Continued*

(c)   Under the same flowrate conditions in the same thickener, the feed concentration increases to 0.24. Estimate the new solids concentration in the overflow and the underflow once steady state has been reached.

(Answers: (a) 0.21; (b) $C_V = 0$, $C_T = 0$, $C_L = 0.29$, $C_B = 0.087$; (c) $C_V = 0.08$, $C_L = 0.34$)

**2.12**
(a)   Using the batch flux plot data given in Table 2.E12.1, graphically determine the limiting feed concentration for a thickener of area 300 m² handling a feed rate of 0.03 m³/s and with an underflow rate of 0.015 m³/s. Determine the underflow concentration and overflow concentration under these conditions. Sketch a possible concentration profile in the thickener clearly indicating the positions of the overflow launder, the feed well and the point of underflow withdrawal (neglect the conical base of the thickener).

**Table 2.E12.1**   Batch flux test data for question 12

| $C$ | 0.01 | 0.02 | 0.04 | 0.06 | 0.08 | 0.10 | 0.12 | 0.14 | 0.16 | 0.18 | 0.20 |
|---|---|---|---|---|---|---|---|---|---|---|---|
| Flux mm/s ($\times 10^3$) | 5.0 | 9.1 | 13.6 | 15.7 | 16.4 | 16.4 | 15.7 | 13.3 | 10.0 | 8.3 | 7.3 |
| $C$ | 0.22 | 0.24 | 0.26 | 0.28 | 0.30 | 0.32 | 0.34 | 0.36 | 0.38 | 0.40 | |
| Flux mm/s ($\times 10^3$) | 7.7 | 5.6 | 5.1 | 4.5 | 4.2 | 3.8 | 3.5 | 3.3 | 3.0 | 2.9 | |

(b)   Under the same flow conditions as above, the concentration in the feed increases to 110% of the limiting value. Estimate the solids concentration in the overflow, in the underflow, in the section of the thickener above the feed well and in the section below the feed well.

(Answers: (a) $C_{Fcrit} = 0.17$; $C_B = 0.05$, $C_B = 0.19$ [two possible values]; $C_L = 0.34$; (b) $C_V = 0.034$; $C_L = 0.34$; $C_T = 0.19$; $C_B = 0.19$)

# 3

# Particle Size Analysis

## 3.1  INTRODUCTION

In many powder handling and processing operations particle size and size distribution play a key role in determining the bulk properties of the powder. Describing the size distribution of the particles making up a powder is therefore central in characterising the powder. In many industrial applications a single number will be required to characterise the particle size of the powder. This can only be done accurately and easily with a mono-sized distribution of spheres or cubes. Real particles with shapes that require more than one dimension to fully describe them and real powders with particles in a range of sizes, mean that in practice the identification of single number to adequately describe the size of the particles is far from straightforward. This chapter deals with how this is done.

## 3.2  DESCRIBING THE SIZE OF A SINGLE PARTICLE

Regular-shaped particles can be accurately described by giving the shape and a number of dimensions. Examples are given in Table 3.1.

The description of the shapes of irregular-shaped particles is a branch of science in itself and will not be covered in detail here. Readers wishing to know more on this topic are referred to Hawkins (1993). However, it will be clear to the reader that no single physical dimension can adequately describe the size of an irregularly shaped particle, just as a single dimension cannot describe the shape of a cylinder, a cuboid or a cone. Which dimension we do use will in practice depend on (a) what property or dimension of the particle we are able to measure and (b) the use to which the dimension is to be put.

**Table 3.1**  Regular-shaped particles

| Shape | Sphere | Cube | Cylinder | Cuboid | Cone |
|---|---|---|---|---|---|
| Dimensions | Radius | Side length | Radius and height | Three side lengths | Radius and height |

If we are using a microscope, perhaps coupled with an image analyser, to view the particles and measure their size, we are looking at a projection of the shape of the particles. Some common diameters used in microscope analysis are statistical diameters such as Martin's diameter (length of the line which bisects the particle image), Feret's diameter (distance between two tangents on opposite sides of the particle) and shear diameter (particle width obtained using an image shearing device) and equivalent circle diameters such as the projected area diameter (area of circle with same area as the projected area of the particle resting in a stable position). Some of these diameters are described in Figure 3.1. We must remember that the orientation of the particle on the microscope slide will affect the projected image and consequently the measured equivalent sphere diameter.

If we use a sieve to measure the particle size we come up with an equivalent

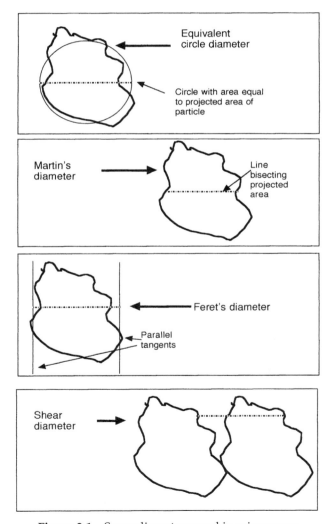

**Figure 3.1**   Some diameters used in microscopy

sphere diameter which is the diameter of a sphere passing through the same sieve aperture. If we use a sedimentation technique to measure particle size then is expressed as the diameter of a sphere having the same sedimentation velocity under the same conditions. Other examples of the properties of particles measured and the resulting equivalent sphere diameters are given in Figure 3.2.

Table 3.2 compares values of these different equivalent sphere diameters used to describe a cuboid of side lengths 1, 3, 5 and a cylinder of diameter 3 and length 1.

The volume equivalent sphere diameter or equivalent volume sphere diameter is a commonly used equivalent sphere diameter. We will see later in the chapter that it is used in the Coulter counter size measurements technique. By definition, the equivalent volume sphere diameter is the diameter of a sphere having the same volume as the particle. The surface-volume diameter is the one measured when we use permeametry (see Section 3.8.4) to measure size. The surface-volume (equivalent sphere) diameter is the diameter of a sphere having the same surface to volume ratio as the particle. *In practice it is important to use the method of size measurement which directly gives the particle size which is relevant to the situation or process of interest.* (See Worked Example 3.1.)

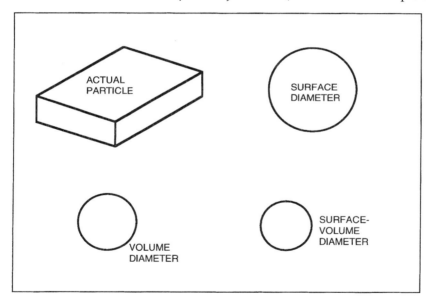

**Figure 3.2**    Comparison of equivalent sphere diameters

**Table 3.2**    Comparison of equivalent sphere diameters

| Shape | Sphere passing the same sieve aperture, $x_p$ | Sphere having the same volume, $x_v$ | Sphere having the same surface area, $x_s$ | Sphere having the same surface to volume ratio, $x_{sv}$ |
|---|---|---|---|---|
| Cuboid | 3 | 3.06 | 3.83 | 1.95 |
| Cylinder | 3 | 2.38 | 2.74 | 1.80 |

## 3.3   DESCRIPTION OF POPULATIONS OF PARTICLES

A population of particles is described by a particle size distribution. Particle size
distributions may be expressed as frequency distribution curves or cumulative
curves. These are illustrated in Figure 3.3. The two are related mathematically
in that the cumulative distribution is the integral of the frequency distribution;
i.e. if the cumulative distribution is denoted as $F$, then the frequency distribution $dF/dx$. For simplicity, $dF/dx$ is often written as $f(x)$. The distributions can
be by number, surface, mass or volume (where particle density does not vary
with size, the mass distribution is the same as the volume distribution).
Incorporating this information into the notation, $f_N(x)$ is the frequency distribution by number, $f_S(x)$ is the frequency distribution by surface, $F_S$ is the
cumulative distribution by surface and $F_M$ is the cumulative distribution by
mass. In reality these distributions are smooth continuous curves. However,
size measurement methods often divide the size spectrum into size ranges or
classes and the size distribution becomes a histogram.

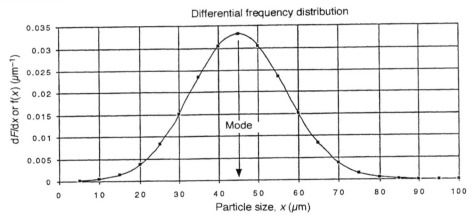

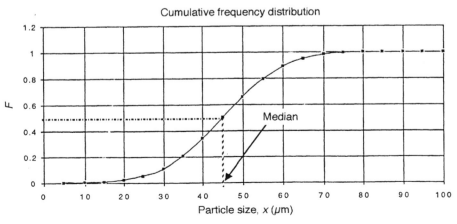

**Figure 3.3**   Typical differential and cumulative frequency distributions

For a given population of particles, the distributions by mass, number and surface can differ dramatically, as can be seen in Figure 3.4.

A further example of difference between distributions for the same population is given in Table 3.3 showing size distributions of man-made objects orbiting the earth (*New Scientist*, 13 October, 1991).

The number distribution tells us that only 0.2% of the objects are greater than 10 cm. However, these larger objects make up 99.96% of the mass of the population, and the 99.3% of the objects which are less than 1.0 cm in size make up only 0.01% of the mass distribution. Which distribution we would use is dependent on the end use of the information.

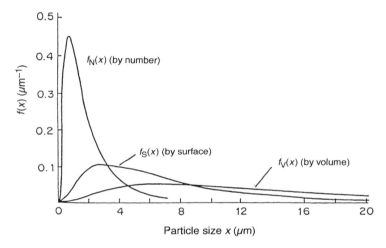

**Figure 3.4**   Comparison between distributions

**Table 3.3**   Mass and number distributions for man-made objects orbiting the earth

| Size (cm) | Number of objects | % by number | % by mass |
|-----------|-------------------|-------------|-----------|
| 10–1000   | 7000              | 0.2         | 99.96     |
| 1–10      | 17 500            | 0.5         | 0.03      |
| 0.1–1.0   | 3 500 000         | 99.3        | 0.01      |
| Total     | 3 524 500         | 100.00      | 100.00    |

## 3.4   CONVERSION BETWEEN DISTRIBUTIONS

Many modern size analysis instruments actually measure a number distribution, which is rarely needed in practice. These instruments include software to convert the measured distribution into more practical distributions by mass, surface etc.

Relating the size distributions by number, $f_N(x)$, and by surface, $f_S(x)$ for a population of particles having the same geometric shape but different size:

Fraction of particles in the size range

$$x \text{ to } x + dx = f_N(x) \, dx$$

Fraction of the total surface of particles in the size range

$$x \text{ to } x + dx = f_S(x) \, dx$$

If N is the total number of particles in the population, number of particles in the size range $x$ to $x + dx = Nf_N(x) \, dx$ and the surface area of these particles $= (x^2 a_S)Nf_N(x) \, dx$ where $a_s$ is the factor relating the linear dimension of the particle to its surface area.

Therefore, the fraction of the total surface area contained on these particles $(f_s(x) \, dx)$ is:

$$\frac{(x^2 a_S)Nf_N(x) \, dx}{S}$$

where S is the total surface area of the population of particles.

For a given population of particles, the total number of particles, N, and the total surface area, S are constant. Also, assuming particle shape is independent of size, $a_S$ is constant, and so

$$f_S(x) \propto x^2 f_N(x) \qquad \text{or} \qquad f_S(x) = k_S x^2 f_N(x) \tag{3.1}$$

where,

$$k_S = \frac{a_S N}{S}$$

Similarly, for the distribution by volume,

$$f_V(x) = k_V x^3 f_N(x) \tag{3.2}$$

where,

$$k_V = \frac{a_V N}{V},$$

where V is the total volume of the population of particles and $a_V$ is the factor relating the linear dimension of the particle to its volume.

And for the distribution by mass,

$$f_m(x) = k_m \rho_p x^3 f_N(x) \tag{3.3}$$

where,

$$k_m = \frac{a_V \rho_p N}{V},$$

assuming particle density $\rho_p$ is independent of size.

The constants $k_S$, $k_V$ and $k_m$ may be found by using the fact that:

$$\int_0^\infty f(x) \, dx = 1 \tag{3.4}$$

Thus, when we convert between distributions it is necessary to make assumptions about the constancy of shape and density with size. Since these assumptions may not be valid, the conversions are likely to be in error. Also, calculation errors are introduced into the conversions. For example, imagine that we used an electron microscope to produce a number distribution of size with a measurement error of $\pm 3\%$. Converting the number distribution to a mass distribution we cube the error involved (i.e. the error becomes $\pm 27\%$). For these reasons, conversions between distributions are to be avoided wherever possible. This can be done by choosing the measurement method which gives the required distribution directly.

## 3.5  DESCRIBING THE POPULATION BY A SINGLE NUMBER

In most practical applications, we require to describe the particle size of a population of particles (millions of them) by a single number. There are many options available; the mode, the median, and several different means including arithmetic, geometric, quadratic, harmonic etc. Whichever expression of central tendency of the particle size of the population we use must reflect the property or properties of the population of importance to us. We are, in fact, modelling the real population with an artificial population of mono-sized particles. This section deals with calculation of the different expressions of central tendency and selection of the appropriate expression for a particular application.

The *mode* is the most frequently occurring size in the sample. We note, however, that for the same sample, different modes would be obtained for distributions by number, surface and volume. The mode has no practical significance as a measure of central tendency and so is rarely used in practice.

The *median* is easily read from the cumulative distribution as the 50% size; the size which splits the distribution into two equal parts. In a mass distribution, for example, half of the particles by mass are smaller than the median size. Since the median is easily determined, it is often used. However, it has no special significance as a measure of central tendency of particle size.

Many different *means* can be defined for a given size distribution; as pointed out by Svarovsky (1990). However, they can all be described by the equation

$$g(\bar{x}) = \frac{\int_0^1 g(x)\,dF}{\int_0^1 dF} \qquad \text{but} \quad \int_0^1 dF = 1 \qquad \text{and so} \quad g(\bar{x}) = \int_0^1 g(x)\,dF \qquad (3.5)$$

where $\bar{x}$ is the mean and $g$ is the weighting function, which is different for each mean definition. Examples are given in Table 3.4.

Equation (3.5) tells us that the mean is the area between the curve and the $F(x)$ axis in a plot of $F(x)$ versus the weighting function $g(x)$ (Figure 3.5). In fact, graphical determination of the mean is always recommended because the distribution is more accurately represented as a continuous curve.

Each mean can be shown to conserve two properties of the original popula-

**Table 3.4**  Definitions of means

| $g(x)$ | Mean and notation |
|--------|-------------------|
| $x$ | arithmetic mean, $\overline{x}_a$ |
| $x^2$ | quadratic mean, $\overline{x}_q$ |
| $x^3$ | cubic mean, $\overline{x}_c$ |
| $\log x$ | geometric mean, $\overline{x}_g$ |
| $1/x$ | harmonic mean, $\overline{x}_h$ |

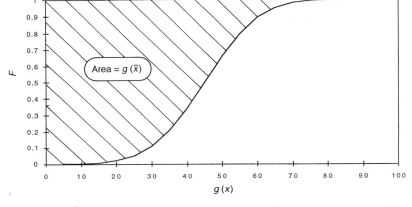

**Figure 3.5**  Plot of cumulative frequency against weighting function $g(x)$. Shaded area is $g(\overline{x}) = \int_0^1 g(x)\,\mathrm{d}F$

tion of particles. For example, the arithmetic mean of the surface distribution conserves the surface and volume of the original population. This is demonstrated in Worked Example 3.3. This mean is commonly referred to as the surface–volume mean or the Sauter mean. The arithmetic mean of the number distribution $\overline{x}_{aN}$ conserves the number and length of the original population and is known as the number-length mean $\overline{x}_{NL}$:

$$\text{number-length mean, } \overline{x}_{NL} = \overline{x}_{aN} = \frac{\displaystyle\int_0^1 x\,\mathrm{d}F_N}{\displaystyle\int_0^1 \mathrm{d}F_N} \tag{3.6}$$

As another example, the quadratic mean of the number distribution $\overline{x}_{qN}$ conserves the number and surface of the original population and is known as the number-surface mean $\overline{x}_{NS}$:

$$\text{number-surface mean, } \overline{x}_{NS}^2 = \overline{x}_{qN}^2 = \frac{\displaystyle\int_0^1 x^2\,\mathrm{d}F_N}{\displaystyle\int_0^1 \mathrm{d}F_N} \tag{3.7}$$

A comparison of the values of the different means and the mode and median for a given particle size distribution is given in Figure 3.6. This figure high-lights two points: (a) that the values of the different expressions of central tendency can vary significantly and (b) that two quite different distributions could have the same arithmetic mean or median etc. If we select the wrong one for our design correlation or quality control we may be in serious error.

So how do we decide which mean particle size is the most appropriate one for a given application? Worked Examples 3.3 and 3.4 indicate how this is done.

## 3.6 EQUIVALENCE OF MEANS

For Equation (3.8), please see Worked Example 3.3 on page 72.

Means of different distributions can be equivalent. For example, as is shown below, the arithmetic mean of a surface distribution is equivalent (numerically equal to) the harmonic mean of a volume (or mass) distribution.

$$\text{arithmetic mean of a surface distribution, } \bar{x}_{aS} = \frac{\int_0^1 x \, dF_s}{\int_0^1 dF_s} \qquad (3.9)$$

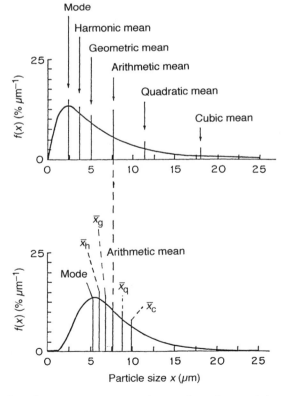

**Figure 3.6** Comparison between measures of central tendency. Adapted from Rhodes 1990. Reproduced by permission

The harmonic mean $\bar{x}_{hV}$ of a volume distribution is defined as

$$\frac{1}{\bar{x}_{hV}} = \frac{\int_0^1 \left(\frac{1}{x}\right) dF_v}{\int_0^1 dF_v} \tag{3.10}$$

From Equations (3.1) and (3.2), the relationship between surface and volume distributions is

$$dF_v = x \, dF_s \frac{k_v}{k_s} \tag{3.11}$$

hence

$$\frac{1}{\bar{x}_{hV}} = \frac{\int_0^1 \left(\frac{1}{x}\right) x \frac{k_v}{k_s} dF_s}{\int_0^1 x \frac{k_v}{k_s} dF_s} = \frac{\int_0^1 dF_s}{\int_0^1 x \, dF_s} \tag{3.12}$$

(assuming $k_s$ and $k_v$ do not vary with size) and so

$$\bar{x}_{hV} = \frac{\int_0^1 x \, dF_s}{\int_0^1 dF_s}$$

which, by inspection, can be seen to be equivalent to the arithmetic mean of the surface distribution $\bar{x}_{aS}$ (Equation (3.9)).

Recalling that $dF_s = x^2 k_s dF_N$, we see from Equation (3.9) that

$$\bar{x}_{aS} = \frac{\int_0^1 x^3 dF_N}{\int_0^1 x^2 dF_N},$$

which is the surface-volume mean, $\bar{x}_{SV}$ (Equation (3.8)).

Summarizing, then, the surface-volume mean may be calculated as the arithmetic mean of the surface distribution or the harmonic mean of the volume distribution. The practical significance of the equivalence of means is that it permits useful means to be calculated easily from a single size analysis.

The reader is invited to investigate the equivalence of other means.

## 3.7   COMMON METHODS OF DISPLAYING SIZE DISTRIBUTIONS

### 3.7.1   Arithmetic-normal Distribution

In this distribution, shown in Figure 3.7, particle sizes with equal differences

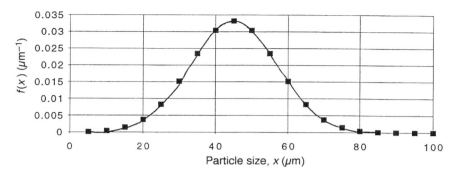

**Figure 3.7** Arithmetic-normal distribution with an arithmetic mean of 45 and standard deviation of 12

from the arithmetic mean occur with equal frequency. Mode, median and arithmetic mean coincide. The distribution can be expressed mathematically by the equation:

$$\frac{dF}{dx} = \frac{1}{\sigma\sqrt{2\pi}} \exp\left[-\frac{(x-\bar{x})^2}{2\sigma^2}\right] \tag{3.13}$$

where, $\sigma$ the standard deviation.

To check for a arithmetic-normal distribution, size analysis data is plotted on normal probability graph paper. On such graph paper a straight line will result if the data fits an arithmetic-normal distribution.

## 3.7.2 Log-normal Distribution

This distribution is more common for naturally occurring particle populations. An example is shown in Figure 3.8. If plotted as $dF/d(\log x)$ versus $x$, rather than $dF/dx$ versus $x$, an arithmetic-normal distribution in $\log x$ results (Figure 3.9). The mathematical expression describing this distribution is

$$\frac{dF}{dz} = \frac{1}{\sigma_z\sqrt{2\pi}} \exp\left[-\frac{(z-\bar{z})^2}{2\sigma_z^2}\right] \tag{3.14}$$

where $z = \log x$, $\bar{z}$ is the arithmetic mean of $\log x$ and $\sigma_z$ is the standard deviation of $\log x$.

To check for a log-normal distribution, size analysis data is plotted on log-normal probability graph paper. Using such graph paper, a straight line will result if the data fits a log-normal distribution.

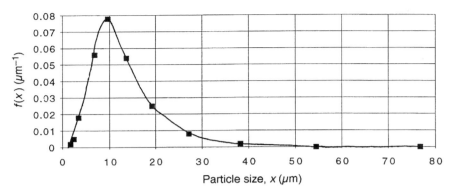

**Figure 3.8**   Log-normal distribution plotted on linear coordinates

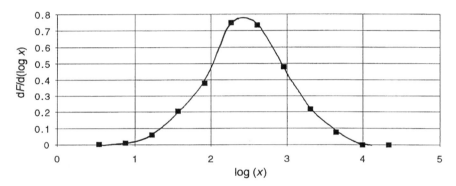

**Figure 3.9**   Log-normal distribution plotted on logarithmic coordinates

## 3.8   METHODS OF PARTICLE SIZE MEASUREMENT

### 3.8.1   Sieving

Dry sieving using woven wire sieves is a simple, cheap method of size analysis suitable for particle sizes greater than 45 μm. Sieving gives a mass distribution and a size known as the sieve diameter. Since the length of the particle does not hinder its passage through the sieve apertures (unless the particle is extremely elongated), the sieve diameter is dependent on the maximum width and maximum thickness of the particle. The most common modern sieves are in sizes such that the ratio of adjacent sieve sizes is the fourth root of two (eg. 45, 53, 63, 75, 90, 107 μm). If standard procedures are followed and care is taken, sieving gives reliable and reproducible size analysis. Air jet sieving, in which the powder on the sieve is fluidized by a jet or air, can achieve analysis down to 20 μm. Analysis down to 5 μm can be achieved by wet sieving, in which the powder sample is suspended in a liquid.

## 3.8.2 Microscopy

The optical microscope may be used to measure particle sizes down to 5 μm. For particles smaller than this diffraction causes the edges of the particle to be blurred and this gives rise to an apparent size. The electron microscope may be used for size analysis below 5 μm. Coupled with an image analysis system the optical microscope or electron microscope can readily give number distributions of size and shape. Such systems calculate various diameters from the projected image of the particles (eg. Martin's, Feret's, shear, projected area diameters etc.). Note that for irregular-shaped particles, the projected area offered to the viewer can vary significantly depending on the orientation of the particle. Techniques such as applying adhesive to the microscope slide may be used to ensure that the particles are randomly orientated.

## 3.8.3 Sedimentation

In this method, the rate of sedimentation of a sample of particles in a liquid is followed. The suspension is dilute and so the particles are assumed to fall at their single particle terminal velocity in the liquid (usually water). Stokes' law is assumed to apply ($Re_p < 0.3$) and so the method using water is suitable only for particles typically less than 50 μm in diameter. The rate of sedimentation of the particles is followed by plotting the suspension density at a certain vertical position against time. The suspension density is directly related to the cumulative undersize and the time is related to the particle diameter via the terminal velocity. This is demonstrated in the following:

Referring to Figure 3.10, the suspension density is sampled at a vertical distance, $h$ below the surface of the suspension. The following assumptions are made:

- The suspension is sufficiently dilute for the particles to settle as individuals (i.e. not hindered settling – see Chapter 2)

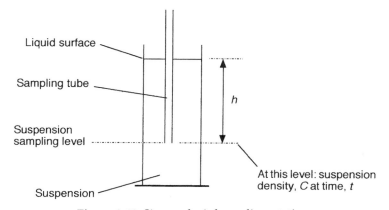

**Figure 3.10** Size analysis by sedimentation

- Motion of the particles in the liquid obeys Stokes' law (true for particles typically smaller than 50 μm)

- Particles are assumed to accelerate rapidly to their terminal free fall velocity $U_T$ so that the time for acceleration is negligible.

Let the original uniform suspension density be $C_0$. Let the suspension density at the sampling point be $C$ at time $t$ after the start of settling. At time $t$ all those particles travelling faster than $h/t$ will have fallen below the sampling point. The sample at time $t$ will therefore consist only of particles travelling a velocity $\leqslant h/t$. Thus, if $C_0$ is representative of the suspension density for the whole population, then $C$ represents the suspension density for all particles which travel at a velocity $\leqslant h/t$, and so $C/C_0$ is the mass fraction of the original particles which travel at a velocity $\leqslant h/t$. That is,

$$\text{cumulative mass fraction} = \frac{C}{C_0}$$

All particles travel at their terminal velocity given by Stokes' law (Chapter 1, Equation (1.13)):

$$U_T = \frac{x^2(\rho_p - \rho_f)g}{18\mu}$$

Thus, equating $U_T$ with $h/t$, we determine the diameter of the particle travelling at our cut-off velocity $h/t$. That is,

$$x = \left( \frac{18\mu h}{t(\rho_p - \rho_f)g} \right)^{1/2} \tag{3.15}$$

Particles smaller than $x$ will travel slower than $h/t$ and will still be in suspension at the sampling point. Corresponding values of $C/C_0$ and $x$ therefore give us the cumulative mass distribution. The particle size measured is the Stokes' diameter, i.e. the diameter of a sphere having the same terminal settling velocity in the Stokes region as the actual particle.

A common form of this method is the Andreason pipette which is capable of measuring in the range 2–100 μm. At size below 2 μm, Brownian motion causes significant errors. Increasing the body force acting on the particles by centrifuging the suspension permits the effects of Brownian motion to be reduced so that particle sizes down to 0.01 μm can be measured. Such a device is known as a pipette centrifuge.

The labour involved in this method may be reduced by using either light absorption or X-ray absorption to measure the suspension density. The light absorption method gives rise to a distribution by surface, whereas the X-ray absorption method gives a mass distribution.

### 3.8.4 Permeametry

This is a method of size analysis based on fluid flow through a packed bed (see Chapter 4). The Carman–Kozeny equation for laminar flow through a randomly packed bed of uniformly sized spheres of diameter $x$ is (Equation 4.9)

$$\frac{(-\Delta p)}{H} = 180\frac{(1-\varepsilon)^2}{\varepsilon^3}\frac{\mu U}{x^2}$$

Where $(-\Delta p)$ is the pressure drop across the bed, $\varepsilon$ is the packed bed void fraction, $H$ is the depth of the bed, $\mu$ is the fluid viscosity and $U$ is the superficial fluid velocity. In Worked Example 3.3, we will see that, when we are dealing with non-spherical particles with a distribution of sizes, the appropriate mean diameter for this equation is the surface-volume diameter $\bar{x}_{SV}$, which may be calculated as the arithmetic mean of the surface distribution, $\bar{x}_{aS}$.

In this method, the pressure gradient across a packed bed of known voidage is measured as a function of flow rate. The diameter we calculate from the Carman–Kozeny equation is the arithmetic mean of the surface distribution (see Worked Example 4.1 in Chapter 4).

### 3.8.5 Electrozone Sensing

Particles are held in supension in a dilute electrolyte which is drawn through a tiny orifice with a voltage applied across it (Figure 3.11). As particles flow through the orifice a voltage pulse is recorded.

The amplitude of the pulse can be related to the volume of the particle passing the orifice. Thus, by electronically counting and classifying the pulses according to amplitude this technique can give a number distribution of the equivalent volume sphere diameter. The lower size limit is dictated by the smallest practical orifice and the upper limit is governed by the need to maintain particles in suspension. Although liquids more viscous than water may be used

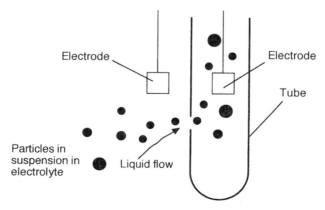

**Figure 3.11**   Schematic of electrozone sensing apparatus

to reduce sedimentation, the practical range of size for this method is 0.3–1000 µm. The most common form of this apparatus is the Coulter counter.

### 3.8.6   Laser Diffraction

This method relies on the fact that for light passing through a suspension, the diffraction angle is inversely proportional to the particle size. An instrument would consist of a laser as a source of coherent light of known fixed wavelength (typically 0.63 µm), a suitable detector (usually a slice of photosensitive silicon with a number of discrete detectors, and some means of passing the sample of particles through the laser light beam (techniques are available for suspending particles in both liquids and gases are drawing them through the beam).

To relate diffraction angle with particle size, early instruments used the Fraunhofer theory, which can give rise to large errors under some circumstances (e.g. when the refractive indices of the particle material and suspending medium approach each other). Modern instruments use the Mie theory for interaction of light with matter. This allows particle sizing in the range 0.1–2000 µm, provided that the refractive indices of the particle material and suspending medium are known.

This method gives a volume distribution and measures a diameter known as the laser diameter. Particle size analysis by laser diffraction is very common in industry today. The associated software permits display of a variety of size distributions and means derived from the original measured distribution.

## 3.9   SAMPLING

In practice, the size distribution of many tonnes of powder are often assumed from an analysis performed on just a few grams or milligrams of sample. The importance of that sample being representative of the bulk powder cannot be overstated. However, as pointed out in Chapter 9 on mixing and segregation, most powder handling and processing operations (pouring, belt conveying, handling in bags or drums, motion of the sample bottle etc.) cause particles to segregate according to size and to a lesser extent density and shape. This natural tendency to segregation means that extreme care must be taken in sampling.

There are two golden rules of sampling:

1. The powder should be in motion when sampled.

2. The whole of the moving stream should be taken for many short time increments.

Since the eventual sample size used in the analysis may be very small, it is often necessary to split the original sample in order to achieve the desired amount for analysis. These sampling rules must be applied at every step of sampling and sample splitting.

Detailed description of the many devices and techniques used for sampling in different process situations and sample dividing are outside the scope of this chapter. However, Allen (1990) gives an excellent account, to which the reader is referred.

## 3.10   WORKED EXAMPLES

### WORKED EXAMPLE 3.1

Calculate the equivalent volume sphere diameter $x_v$ and the surface-volume equivalent sphere diameter $x_{sv}$ of a cuboid particle of side length 1, 2, 4 mm.

*Solution*

The volume of cuboid $= 1 \times 2 \times 4 = 8$ mm$^3$.
The surface area of the particle $= (1 \times 2) + (1 \times 2) + (1 + 2 + 1 + 2) \times 4 = 28$ mm$^2$.
The volume of sphere of diameter $x_v$ is $\pi x_v^3/6$
Hence, diameter of a sphere having a volume of 8 mm$^3$, $x_v = 2.481$ mm.
The *equivalent volume sphere diameter* $x_v$ of the cuboid particle is therefore $x_v = 2.481$ mm
The surface to volume ratio of the cuboid particle $= \frac{28}{8} = 3.5$ mm$^2$/mm$^3$.
The surface to volume ratio for a sphere of diameter $x_{sv}$ is therefore $6/x_{sv}$
Hence, the diameter of a sphere having the same surface to volume ratio as the particle $= 6/3.5 = 1.714$ mm.
The *surface-volume equivalent sphere diameter* of the cuboid, $x_{sv} = 1.714$ mm

### WORKED EXAMPLE 3.2

Convert the surface distribution described by the following equation to a cumulative volume distribution:

$$F_S = (x/45)^2 \quad \text{for } x \leqslant 45 \text{ μm}$$

$$F_S = 1 \qquad\qquad \text{for } x > 45 \text{ μm}$$

*Solution*

From Equations (3.1)–(3.3),

$$f_v(x) = \frac{k_v}{k_s} x f_s(x)$$

Integrating between sizes 0 and $x$:

$$F_v(x) = \int_0^x \left(\frac{k_v}{k_s}\right) x f_s(x)\, dx$$

Noting that $f_s(x) = dF_s/dx$, we see that

$$f_s(x) = \frac{\mathrm{d}}{\mathrm{d}x}\left(\frac{x}{45}\right)^2 = \frac{2x}{(45)^2}$$

and our integral becomes

$$F_v(x) = \int_0^x \left(\frac{k_v}{k_s}\right)\frac{2x^2}{(45)^2}\,\mathrm{d}x$$

Assuming that $k_v$ and $k_s$ are independent of size,

$$F_v(x) = \left(\frac{k_v}{k_s}\right)\int_0^x \frac{2x^2}{(45)^2}\,\mathrm{d}x$$

$$= \frac{2}{3}\left(\frac{x^3}{(45)^2}\right)\frac{k_v}{k_s}$$

$k_v/k_s$ may be found by noting that $F_v(45) = 1$; hence

$$\frac{90}{3}\frac{k_v}{k_s} = 1 \quad \text{and so} \quad \frac{k_v}{k_s} = 0.0333$$

Thus, the formula for the volume distribution is

$$F_v = 1.096 \times 10^{-5}x^3 \quad \text{for } x \leqslant 45\ \mu m$$

$$F_v = 1 \qquad\qquad\qquad \text{for } x > 45\ \mu m$$

## WORKED EXAMPLE 3.3

What mean particle size do we use in calculating the pressure gradient for flow of a fluid through a packed bed of particles using the Carman–Kozeny equation (see Chapter 4)?

### *Solution*

The Carman–Kozeny equation for laminar flow through a randomly packed bed of particles is

$$\frac{(-\Delta p)}{L} = K\frac{(1-\varepsilon)^2}{\varepsilon^3}S_v^2\mu U$$

where $S_v$ is the specific surface area of the bed of particles (particle surface area per unit particle volume) and the other terms are defined in Chapter 4. If we assume that the bed voidage is independent of particle size, then to write the equation in terms of a mean particle size, we must express the specific surface, $S_v$, in terms of that mean. The particle size we use must give the same value of $S_v$ as the original population or particles. Thus the mean diameter $\bar{x}$, must conserve the surface and volume of the population; that is the mean must enable us to calculate the total volume from the total surface of the particles. This mean is the surface-volume mean $\bar{x}_{sv}$

$$\bar{x}_{sv} \times \text{(total surface)} \times \frac{a_v}{a_s} = \text{(total volume)} \left( \text{eg. for a spheres,} \frac{a_v}{a_s} = 1/6 \right)$$

$$\text{and therefore } \bar{x}_{sv} \int_0^\infty f_s(x)\,\mathrm{d}x \cdot \frac{k_v}{k_s} = \int_0^\infty f_v(x)\,\mathrm{d}x$$

$$\text{Total volume of particles, } V = \int_0^\infty x^3 a_v N f_N(x)\,\mathrm{d}x$$

$$\text{Total surface area of particles, } S = \int_0^\infty x^2 a_s N f_N(x)\,\mathrm{d}x$$

$$\text{Hence, } \bar{x}_{sv} = \frac{a_s}{a_v} \frac{\int_0^\infty x^3 a_v N f_N(x)\,\mathrm{d}x}{\int_0^\infty x^2 a_s N f_N(x)\,\mathrm{d}x}$$

Then, since $a_V$, $a_S$ and N are independent of size, $x$,

$$\bar{x}_{sv} = \frac{\int_0^\infty x^3 f_N(x)\,\mathrm{d}x}{\int_0^\infty x^2 f_N(x)\,\mathrm{d}x} = \frac{\int_0^1 x^3 \mathrm{d}F_N}{\int_0^1 x^2 \mathrm{d}F_N}$$

This is the definition of the mean which conserves surface and volume, known as the surface-volume mean, $\bar{x}_{SV}$.
So,

$$\bar{x}_{SV} = \frac{\int_0^1 x^3 \mathrm{d}F_N}{\int_0^1 x^2 \mathrm{d}F_N} \tag{3.8}$$

The correct mean particle diameter is therefore the surface-volume mean as defined above. (We saw in Section 3.6 that this may be calculated as the arithmetic mean of the surface distribution $\bar{x}_{aS}$, or the harmonic mean of the volume distribution.) Then in the Carman–Kozeny equation we make the following substitution for $S_v$:

$$S_v = \frac{1}{\bar{x}_{SV}} \frac{k_s}{k_v}$$

e.g. for spheres, $S_v = 6/\bar{x}_{SV}$

## WORKED EXAMPLE 3.4 (AFTER SVAROVSKY, 1990)

A gravity settling device processing a feed with size distribution $F(x)$ and operates with a grade efficiency $G(x)$. Its total efficiency is defined as

$$E_T = \int_0^1 G(x)\,\mathrm{d}F_M$$

How is the mean particle size to be determined?

## Solution

Assuming plug flow (see Chapter 2), $G(x) = U_T A/Q$ where, $A$ is the settling area, $Q$ is the volume flow rate of suspension and $U_T$ is the single particle terminal velocity for particle size $x$, given by (in the Stokes region):

$$U_T = \frac{x^2(\rho_p - \rho_f)g}{18\mu} \qquad \text{(Chapter 1)}$$

$$\text{hence, } E_T = \frac{Ag(\rho_p - \rho_f)}{18\mu Q} \int_0^1 x^2 dF_M$$

where $\int_0^1 x^2 dF_M$ is seen to be the definition of the quadratic mean of the distribution by mass $\bar{x}_{qM}$ (see Table 3.4).

This approach may be used to determine the correct mean to use in many applications.

## WORKED EXAMPLE 3.5

A Coulter counter analysis of a cracking catalyst sample gives the following cumulative volume distribution:

| Channel | 1 | 2 | 3 | 4 | 5 | 6 | 7 | 8 |
|---|---|---|---|---|---|---|---|---|
| % volume differential | 0 | 0.5 | 1.0 | 1.6 | 2.6 | 3.8 | 5.7 | 8.7 |

| Channel | 9 | 10 | 11 | 12 | 13 | 14 | 15 | 16 |
|---|---|---|---|---|---|---|---|---|
| % volume differential | 14.3 | 22.2 | 33.8 | 51.3 | 72.0 | 90.9 | 99.3 | 100 |

(a)  Plot the cumulative volume distribution versus size and determine the median size.

(b)  Determine the surface distribution, giving assumptions. Compare with the volume distribution.

(c)  Determine the harmonic mean diameter of the volume distribution.

(d)  Determine the arithmetic mean diameter of the surface distribution.

## Solution

With the Coulter counter the channel size range differs depending on the tube in use. We therefore need the additional information that in this case channel 1 covers the size range 3.17 μm to 4.0 μm, channel 2 covers the range 4.0 μm to 5.04 μm and so on up to channel 16, which covers the range 101.4 μm to 128 μm. The ratio of adjacent size range boundaries is always the cube root of 2. For example,

$$\sqrt[3]{2} = \frac{4.0}{3.17} = \frac{5.04}{4.0} = \frac{128}{101.4} \quad \text{etc.}$$

The resulting lower and upper sizes for the channels are shown in columns 2 and 3 of Table 3W5.1.

(a) The cumulative undersize distribution is shown numerically in column 5 of Table 3.5 and graphically in Figure 3W5.1. By inspection, we see that the median size is 50 μm, ie. 50% by volume of the particles is less than 50 μm.

(b) The surface distribution is related to the volume distribution by the expression:

$$f_s(x) = \frac{f_v(x)}{x} \times \frac{k_s}{k_v} \qquad \text{(from Equations (3.1) and (3.2))}$$

Recalling that $f(x) = dF/dx$ and integrating between 0 and $x$:

$$\frac{k_s}{k_v} \int_0^x \frac{1}{x} \frac{dF_v}{dx} \, dx = \int_0^x \frac{dF_s}{dx} \, dx$$

$$\text{or} \quad \frac{k_s}{k_v} \int_0^x \frac{1}{x} dF_v = \int_0^x dF_s = F_s(x)$$

(assuming particle shape is invariant with size so that $k_s/k_v$ is constant).

So the surface distribution can be found from the area under a plot of $1/x$ versus $F_v$ multiplied by the factor $k_s/k_v$ (which is found by noting that $\int_{x=0}^{x=\infty} dF_s = 1$).

Column 7 of Table 3W5.1 shows the area under $1/x$ versus $F_v$. The factor $k_s/k_v$ is therefore equal to 0.0278. Dividing the values of column 7 by 0.0278 gives the surface distribution $F_s$ shown in column 8. The surface distribution is shown graphically in Figure 3W5.2. The shape of the surface distribution is quite different from that of the volume distribution; the smaller particles make up a high proportion of the total surface. The median of the surface distribution is around 35 μm, i.e. particles under 35 μm contribute 50% of the total surface area.

(c) The harmonic mean of the volume distribution is given by

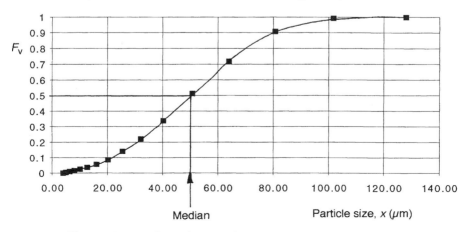

**Figure 3W5.1**  Cumulative volume distribution (Example 3.5)

**Table 3W5.1**  Size distribution data associated with Worked Example 3.5

| 1 Channel number | 2 Lower size of range (micron) | 3 Upper size of range (micron) | 4 Cumulative percent undersize | 5 $F_v$ | 6 $1/x$ | 7 Cumulative area under $F_v$ versus $1/x$ | 8 $F_s$ | 9 Cumulative area under $F_s$ versus $x$ |
|---|---|---|---|---|---|---|---|---|
| 1 | 3.17 | 4.00 | 0 | 0 | 0.2500 | 0.0000 | 0.0000 | 0.0000 |
| 2 | 4.00 | 5.04 | 0.5 | 0.005 | 0.1984 | 0.0011 | 0.0403 | 0.1823 |
| 3 | 5.04 | 6.35 | 1 | 0.01 | 0.1575 | 0.0020 | 0.0723 | 0.3646 |
| 4 | 6.35 | 8.00 | 1.6 | 0.016 | 0.1250 | 0.0029 | 0.1028 | 0.5834 |
| 5 | 8.00 | 10.08 | 2.6 | 0.026 | 0.0992 | 0.0040 | 0.1432 | 0.9480 |
| 6 | 10.08 | 12.70 | 3.8 | 0.038 | 0.0787 | 0.0050 | 0.1816 | 1.3855 |
| 7 | 12.70 | 16.00 | 5.7 | 0.057 | 0.0625 | 0.0064 | 0.2299 | 2.0782 |
| 8 | 16.00 | 20.16 | 8.7 | 0.087 | 0.0496 | 0.0081 | 0.2904 | 3.1720 |
| 9 | 20.16 | 25.40 | 14.3 | 0.143 | 0.0394 | 0.0106 | 0.3800 | 5.2138 |
| 10 | 25.40 | 32.00 | 22.2 | 0.222 | 0.0313 | 0.0134 | 0.4804 | 8.0942 |
| 11 | 32.00 | 40.32 | 33.8 | 0.338 | 0.0248 | 0.0166 | 0.5973 | 12.3236 |
| 12 | 40.32 | 50.80 | 51.3 | 0.513 | 0.0197 | 0.0205 | 0.7374 | 18.7041 |
| 13 | 50.80 | 64.00 | 72 | 0.72 | 0.0156 | 0.0242 | 0.8689 | 26.2514 |
| 14 | 64.00 | 80.63 | 90.9 | 0.909 | 0.0124 | 0.0268 | 0.9642 | 33.1424 |
| 15 | 80.63 | 101.59 | 99.3 | 0.993 | 0.0098 | 0.0277 | 0.9978 | 36.2051 |
| 16 | 101.59 | 128.00 | 100 | 1 | 0.0078 | **0.0278** | 1.0000 | **36.4603** |

Note: Based on arithmetic means of size ranges

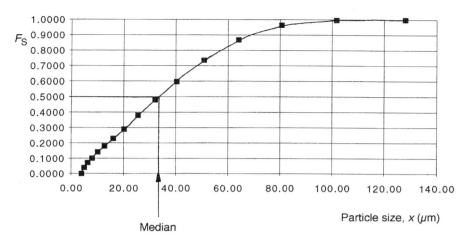

**Figure 3W5.2** Cumulative surface distribution (Example 3.5)

$$\frac{1}{\overline{x}_{hV}} = \int_0^1 \left(\frac{1}{x}\right) dF_v$$

This can be calculated graphically from a plot of $F_v$ versus $1/x$ or numerically from the tabulated data in column 7 of Table 3.W5.1. Hence,

$$\frac{1}{\overline{x}_{hV}} = \int_0^1 \left(\frac{1}{x}\right) dF_v = 0.0278$$

and so, $\overline{x}_{hV} = 36$ μm.

We recall that the harmonic mean of the volume distribution is equivalent to the surface-volume mean of the population.

(d) the arithmetic mean of the surface distribution is given by

$$\overline{x}_{aS} = \int_0^1 x \, dF_s$$

This may be calculated graphically from our plot of $F_s$ versus $x$ (Figure 3W5.2) or numerically using the data in Table 3.W5.1. This area calculation as shown in Column 9 of the table shows the cumulative area under a plot of $F_s$ versus $x$ and so the last figure in this column is equivalent to the above integral. Thus:

$$\overline{x}_{as} = 36.4 \text{ μm}.$$

We may recall that the arithmetic mean of the surface distribution is also equivalent to the surface-volume mean of the population. This value compares well with the value obtained in (c) above.

## WORKED EXAMPLE 3.6

Consider a cuboid particle $5.00 \times 3.00 \times 1.00$ mm. Calculate for this particle the following diameters:

(a) the volume diameter (the diameter of a sphere having the same volume as the particle);

(b) the surface diameter (the diameter of a sphere having the same surface area as the particle);

(c) the surface-volume diameter (the diameter of a sphere having the same external surface to volume ratio as the particle);

(d) the sieve diameter (the width of the minimum aperture through which the particle will pass);

(e) the projected area diameters (the diameter of a circle having the same area as the projected area of the particle resting in a stable position).

### Solution

(a) Volume of the particle $= 5 \times 3 \times 1 = 15 \text{ mm}^3$

Volume of a sphere $= \dfrac{\pi x_v^3}{6}$,

Thus volume diameter, $x_v = \sqrt[3]{\dfrac{15 \times 6}{\pi}} = 3.06 \text{ mm}$

(b) Surface area of the particle $= 2 \times (5 \times 3) + 2 \times (1 \times 3) + 2 \times (1 \times 5) = 46 \text{ mm}^2$

Surface area of sphere $= \pi x_s^2$

Therefore , surface diameter, $x_s = \sqrt{\dfrac{46}{\pi}} = 3.83 \text{ mm}$

(c) Ratio of surface to volume of the particle $= 46/15 = 3.0667$

For a sphere, surface to volume ratio $= \dfrac{6}{x_{sv}}$

Therefore, $x_{sv} = \dfrac{6}{3.0667} = 1.96 \text{ mm}$

(d) The smallest square aperture through which this particle will pass is 3 mm. Hence, the sieve diameter, $x_p = 3 \text{ mm}$.

(e) This particle has three projected areas in stable positions:

$$\text{area } 1 = 3 \text{ mm}^2; \text{ area } 2 = 5 \text{ mm}^2; \text{ area } 3 = 15 \text{ mm}^2$$

$$\text{area of circle } = \dfrac{\pi x^2}{4}$$

hence, projected area diameters:

$$\text{projected area diameter } 1 = 1.95 \text{ mm};$$

$$\text{projected area diameter } 2 = 2.52 \text{ mm};$$

$$\text{projected area diameter } 3 = 4.37 \text{ mm}.$$

## EXERCISES

**3.1** For a regular cuboid particle of dimensions $1.00 \times 2.00 \times 6.00$ mm, calculate the following diameters:

(a) the equivalent volume sphere diameter;

(b) the equivalent surface sphere diameter;

(c) the surface-volume diameter (the diameter of a sphere having the same external surface to volume ratio as the particle).

(d) the sieve diameter (the width of the minimum aperture through which the particle will pass)

(e) the projected area diameters (the diameter of a circle having the same area as the projected area of the particle resting in a stable position).

(Answer: (a) 2.84 mm; (b) 3.57 mm; (c) 1.80 mm; (d) 2.00 mm; (e) 2.76 mm, 1.60 mm and 3.91 mm.)

**3.2** Repeat Exercise 3.1 for a regular cylinder of diameter 0.100 mm and length 1.00 mm.

(Answer: (a) 0.247 mm; (b) 0.324 mm; (c) 0.142 mm; (d) 0.10 mm; (e) 0.10 mm (unlikely to be stable in this position, and 0.357 mm.)

**3.3** Repeat Exercise 3.1 for a disc-shaped particle of diameter 2.00 mm and length 0.500 mm.

(Answer: (a) 1.44 mm; (b) 1.73 mm; (c) 1.00 mm; (d) 2.00 mm; (e) 2.00 mm and 1.13 mm (unlikely to be stable in this position).)

**3.4** 1.28 g of a powder of particle density 2500 $kg/m^3$ are charged into the cell of an apparatus for measurement of particle size and specific surface area by permeametry. The cylindrical cell has a diameter of 1.14 cm and the powder forms a bed of depth 1 cm. Dry air of density 1.2 $kg/m^3$ and viscosity $18.4 \times 10^{-6}$ Pa s flows at a rate of 36 $cm^3/min$ through the powder (in a direction parallel to the axis of the cylindrical cell) and producing a pressure difference of 100 mm of water across the bed. Determine the surface-volume mean diameter and the specific surface of the powder sample.

(Answer: 20 μm; 120 $m^2/kg$)

**3.5** 1.1 g of a powder of particle density 1800 kg/m$^3$ are charged into the cell of an apparatus for measurement of particle size and specific surface area by permeametry. The cylindrical cell has a diameter of 1.14 cm and the powder forms a bed of depth 1 cm. Dry air of density 1.2 kg/m$^3$ and viscosity $18.4 \times 10^{-6}$ Pa s flows through the powder (in a direction parallel to the axis of the cylindrical cell). The measured variation in pressure difference across the bed with changing air flow rate is given below:

| Air flow (cm$^3$/min) | 20 | 30 | 40 | 50 | 60 |
|---|---|---|---|---|---|
| Pressure difference across the bed (mm of water) | 56 | 82 | 112 | 136 | 167 |

Determine the surface-volume mean diameter and the specific surface of the powder sample.

(Answer: 33 µm; 100 m$^2$/kg)

**3.6** Estimate the (a) arithmetic mean, (b) quadratic mean, (c) cubic mean, (d) geometric mean and (e) harmonic mean of the following distribution.

| Size | 2 | 2.8 | 4 | 5.6 | 8 | 11.2 | 16 | 22.4 | 32 | 44.8 | 64 | 89.6 |
|---|---|---|---|---|---|---|---|---|---|---|---|---|
| cumulative % undersize | 0.1 | 0.5 | 2.7 | 9.6 | 23 | 47.9 | 73.8 | 89.8 | 97.1 | 99.2 | 99.8 | 100 |

(Answer: (a) 13.6; (b) 16.1; (c) 19.3; (d) 11.5; (e) 9.8)

**3.7** The following volume distribution was derived from a sieve analysis:

| Size (µm) | 37–45 | 45–53 | 53–63 | 63–75 | 75–90 | 90–106 | 106–126 | 126–150 | 150–180 | 180–212 |
|---|---|---|---|---|---|---|---|---|---|---|
| Volume % in range | 0.4 | 3.1 | 11 | 21.8 | 27.3 | 22 | 10.1 | 3.9 | 0.4 | 0 |

(a) Estimate the arithmetic mean of the volume distribution.

From the volume distribution derive the number distribution and the surface distribution, giving assumptions made.
   Estimate:

(b)  the mode of the surface distribution;

(c)  the harmonic mean of the surface distribution.

Show that the harmonic mean of the surface distribution conserves the surface to volume ratio of the population of particles.

(Answer: (a) 86 µm; (b) 70 µm; (c) 76 µm)

# 4

# Fluid Flow Through a
# Packed Bed of Particles

## 4.1 PRESSURE DROP–FLOW RELATIONSHIP

### 4.1.1 Laminar Flow

In the 19th century Darcy (1856) observed that the flow of water through a packed bed of sand was governed by the relationship:

$$\begin{bmatrix} \text{pressure} \\ \text{gradient} \end{bmatrix} \propto \begin{bmatrix} \text{liquid} \\ \text{velocity} \end{bmatrix} \quad \text{or} \quad \frac{(-\Delta p)}{H} \propto U \tag{4.1}$$

where $U$ is the superficial fluid velocity through the bed and $(-\Delta p)$ is the frictional pressure drop across a bed depth $H$. (Superficial velocity = fluid volumetric flow rate / cross-sectional area of bed, $Q/A$.)

The flow of a fluid through a packed bed of solid particles may be analysed in terms of the fluid flow through tubes. The starting point is the Hagen–Poiseuille equation for laminar flow through a tube:

$$\frac{(-\Delta p)}{H} = \frac{32\mu U}{D^2} \tag{4.2}$$

where $D$ is the tube diameter and $\mu$ is the fluid viscosity.

Consider the packed bed to be equivalent to many tubes of equivalent diameter $D_e$ following tortuous paths of equivalent length $H_e$ and carrying fluid with a velocity $U_i$. Then, from Equation (4.2),

$$\frac{(-\Delta p)}{H_e} = K_1 \frac{\mu U_i}{D_e^2} \tag{4.3}$$

$U_i$ is the actual velocity of fluid through the interstices of the packed bed and is related to superficial fluid velocity by

$$U_i = U/\varepsilon \tag{4.4}$$

where $\varepsilon$ is the voidage or void fraction of the packed bed. (Refer to Section 6.1.4 for discussion on actual and superficial velocities.)

Although the paths of the tubes are tortuous, we can assume that their actual length is proportional to the bed depth, that is,

$$H_e = K_2 H \tag{4.5}$$

The tube equivalent diameter is defined as

$$\frac{\text{flow area}}{\text{wetted perimeter}}$$

where flow area $= \varepsilon A$, where $A$ is the cross-sectional area of the vessel holding the bed;

wetted perimeter $= S_B A$, where $S_B$ is the particle surface area per unit volume of the bed.

That this is so may be demonstrated by comparison with pipe flow:

Total particle surface area in the bed $= S_B A H$. For a pipe,

$$\text{wetted perimeter} = \frac{\text{surface}}{\text{length}}$$

and so for the packed bed wetted perimeter $= \dfrac{S_B A H}{H} = S_B A.$

Now if $S_v$ is the surface area per unit volume of particles, then

$$S_v(1 - \varepsilon) = S_B \tag{4.6}$$

since $\left[\dfrac{\text{surface of particles}}{\text{volume of particles}}\right] \times \left[\dfrac{\text{volume of particles}}{\text{volume of bed}}\right] = \left[\dfrac{\text{surface of particles}}{\text{volume of bed}}\right]$

and so, equivalent diameter, $D_e = \dfrac{\varepsilon A}{S_B A} = \dfrac{\varepsilon}{S_v(1 - \varepsilon)}$ $\tag{4.7}$

Substituting Equations (4.4), (4.5) and (4.7) in (4.3):

$$\frac{(-\Delta p)}{H} = K_3 \frac{(1 - \varepsilon)^2}{\varepsilon^3} \mu U S_v^2 \tag{4.8}$$

where $K_3 = K_1 K_2$. Equation (4.8) is known as the Carman–Kozeny equation (after the work of Carman and Kozeny (Kozeny, 1927, 1933; Carman, 1937)) describing laminar flow through randomly packed particles. The constant $K_3$ depends on particle shape and surface properties and has been found by experiment to have a value of about 5. Taking $K_3 = 5$, for laminar flow through a randomly packed bed of monosized spheres of diameter $x$ (for which $S = 6/x$) the Carman–Kozeny equation becomes

$$\frac{(-\Delta p)}{H} = 180 \frac{\mu U}{x^2} \frac{(1 - \varepsilon)^2}{\varepsilon^3} \tag{4.9}$$

This is the most common form in which the Carman–Kozeny equation is quoted.

### 4.1.2   Turbulent Flow

For turbulent flow through a randomly packed bed of monosized spheres of diameter $x$ the equivalent equation is:

$$\frac{(-\Delta p)}{H} = 1.75 \frac{\rho_f U^2}{x} \frac{(1 - \varepsilon)}{\varepsilon^3} \tag{4.10}$$

### 4.1.3   General Equation for Turbulent and Laminar Flow

Based on extensive experimental data covering a wide range of size and shape of particles, Ergun (1952) suggested the following general equation (4.11) for any flow conditions:

$$\frac{(-\Delta p)}{H} = 150 \frac{\mu U}{x^2} \frac{(1 - \varepsilon)^2}{\varepsilon^3} + 1.75 \frac{\rho_f U^2}{x} \frac{(1 - \varepsilon)}{\varepsilon^3}$$

$$\begin{bmatrix} \text{laminar} \\ \text{component} \end{bmatrix} \qquad \begin{bmatrix} \text{turbulent} \\ \text{component} \end{bmatrix} \tag{4.11}$$

This is known as the Ergun equation for flow through a randomly packed bed of spherical particles of diameter $x$. Ergun's equation additively combines the laminar and turbulent components of the pressure gradient. Under laminar conditions, the first term dominates and the equation reduces to the Carman–Kozeny equation (Equation 4.9), but with the constant 150 rather than 180. (The difference in the values of the constants is probably due to differences in shapes and packing of the particles.) In laminar flow the pressure gradient increases linearly with superficial fluid velocity and independent of fluid density. Under turbulent flow conditions, the second term dominates; the pressure gradient increases as the square of superficial fluid velocity and is independent of fluid viscosity. In terms of the Reynolds number defined in Equation (4.12), fully laminar condition exist for $Re^*$ less than about 10 and fully turbulent flow exists at Reynolds numbers greater than around 2000.

$$Re^* = \frac{x U \rho_f}{\mu (1 - \varepsilon)} \tag{4.12}$$

In practice, the Ergun equation is often used to predict packed bed pressure gradient over the entire range of flow conditions. For simplicity, this practice is followed in the worked examples and questions in this chapter.

Ergun also expressed flow through a packed bed in terms of a friction factor defined in Equation (4.13):

$$\text{friction factor, } f^* = \frac{(-\Delta p)}{H} \frac{x}{\rho_f U^2} \frac{\varepsilon^3}{(1-\varepsilon)} \qquad (4.13)$$

(Compare the form of this friction factor with the familiar Fanning friction factor for flow through pipes.)

Equation (4.11) then becomes

$$f^* = \frac{150}{Re^*} + 1.75 \qquad (4.14)$$

with

$$f^* = \frac{150}{Re^*} \quad \text{for} \quad Re^* < 10; \qquad \text{and} \qquad f^* = 1.75 \quad \text{for} \quad Re^* > 2000$$

(see Figure 4.1).

### 4.1.4  Non-spherical particles

The Ergun and Carman–Kozeny equations also accommodate non-spherical particles if $x$ is replaced by $x_{sv}$ the diameter of a sphere having the same surface to volume ratio as the non-spherical particles in question. Use of $x_{sv}$ gives the correct value of specific surface $S$ (surface area of particles per unit volume of particles). The relevance of this will be apparent if Equation (4.8) is

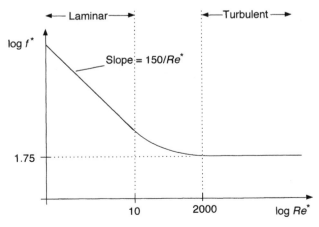

**Figure 4.1**  Friction factor versus Reynolds number plot for fluid flows through a packed bed of spheres

recalled. Thus, in general, the Ergun equation for flow through a randomly packed bed of particles of surface – volume diameter $x_{sv}$ becomes

$$\frac{(-\Delta p)}{H} = 150 \frac{\mu U (1 - \varepsilon)^2}{x_{sv}^2} \frac{1}{\varepsilon^3} + 1.75 \frac{\rho_f U^2 (1 - \varepsilon)}{x_{sv}} \frac{1}{\varepsilon^3} \qquad (4.15)$$

and the Carman–Kozeny equation for laminar flow through a randomly packed bed of particles of surface – volume diameter $x_{sv}$ becomes:

$$\frac{(-\Delta p)}{H} = 180 \frac{\mu U (1 - \varepsilon)^2}{x_{sv}^2} \frac{1}{\varepsilon^3} \qquad (4.16)$$

It was shown if Chapter 1 that if the particles in the bed are not mono-sized, then the correct mean size to use in these equations is the surface – volume mean $\bar{x}_{sv}$.

## 4.2 FILTRATION

### 4.2.1 Introduction

As an example of the application of the above analysis for flow through packed beds of particles, we will briefly consider cake filtration. Cake filtration is widely used in industry to separate solid particles from suspension in liquid. It involves the build up of a bed or 'cake' of particles on a porous surface known as the filter medium, which commonly takes the form of a woven fabric. In cake filtration the pore size of the medium is less than the size of the particles to be filtered. It will be appreciated that this filtration process can be analysed in terms of the flow of fluid through a packed bed of particles, the depth of which is increasing with time. In practice the voidage of the cake may also change with time. However, we will first consider the case where the cake voidage is constant; i.e. an incompressible cake.

### 4.2.2 Incompressible Cake

Firstly, if we ignore the filter medium and consider only the cake itself, the pressure drop versus liquid flow relationship is described by the Ergun equation (Equation (4.15)). The particle size and range of liquid flow and properties commonly used in industry give rise to laminar flow and so the second (turbulent) term vanishes. For a given slurry (particle properties fixed) the resulting cake resistance is defined as

$$\text{cake resistance, } r_c = \frac{150 (1 - \varepsilon)^2}{x_{sv}^2} \frac{1}{\varepsilon^3} \qquad (4.17)$$

and so Equation (4.15) becomes

$$\frac{(-\Delta p)}{H} = r_c \mu U \tag{4.18}$$

If $V$ is the volume of filtrate (liquid) passed in a time $t$ and $dV/dt$ is the instantaneous volumetric flowrate of filtrate at time $t$, then

$$\text{superficial filtrate velocity at time } t, U = \frac{1}{A}\frac{dV}{dt} \tag{4.19}$$

Each unit volume of filtrate is assumed to deposit a certain mass of particles, which form a certain volume of cake. This is expressed as $\phi$, the volume of cake formed by the passage of unit volume of filtrate.

$$\phi = \frac{HA}{V} \tag{4.20}$$

and so Equation (4.18) becomes

$$\frac{dV}{dt} = \frac{A^2(-\Delta p)}{r_c \mu \phi V} \tag{4.21}$$

*Constant rate filtration*

If the filtration rate $dV/dt$ is constant, then the pressure drop across the filter cake will increase in direct proportion to the volume of filtrate passed $V$.

*Constant pressure drop filtration*

If $(-\Delta p)$ is constant then,

$$\frac{dV}{dt} \propto \frac{1}{V}$$

or, integrating Equation (4.21),

$$\frac{t}{V} = C_1 V \tag{4.22}$$

$$\text{where, } C_1 = \frac{r_c \mu \phi}{2A^2(-\Delta p)} \tag{4.23}$$

### 4.2.3   Including the Resistance of the Filter Medium

The total resistance to flow is the sum of the resistance of the cake and the filter medium. Hence,

$$\begin{bmatrix} \text{total pressure} \\ \text{drop} \end{bmatrix} = \begin{bmatrix} \text{pressure drop} \\ \text{across medium} \end{bmatrix} + \begin{bmatrix} \text{pressure drop} \\ \text{across cake} \end{bmatrix}$$

$$(-\Delta p) = (-\Delta p_{\mathrm{m}}) + (-\Delta p_{\mathrm{c}})$$

If the medium is assumed to behave as a packed bed of depth $H_{\mathrm{m}}$ and resistance $r_{\mathrm{m}}$ obeying the Carman–Kozeny equation, then

$$(-\Delta p) = \frac{1}{A}\frac{\mathrm{d}V}{\mathrm{d}t}(r_{\mathrm{m}}\mu H_{\mathrm{m}} + r_{\mathrm{c}}\mu H_{\mathrm{c}}) \tag{4.24}$$

The medium resistance is usually expressed as the equivalent thickness of cake $H_{\mathrm{eq}}$:

$$r_{\mathrm{m}} H_{\mathrm{m}} = r_{\mathrm{c}} H_{\mathrm{eq}}$$

Hence, combining with Equation (4.20),

$$H_{\mathrm{eq}} = \frac{\phi V_{\mathrm{eq}}}{A} \tag{4.25}$$

where $V_{\mathrm{eq}}$ is the volume of filtrate that must pass in order to create a cake of thickness $H_{\mathrm{eq}}$. The volume $V_{\mathrm{eq}}$ depends only on the properties of the suspension and the filter medium.

Equation (4.24) becomes

$$\frac{1}{A}\frac{\mathrm{d}V}{\mathrm{d}t} = \frac{(-\Delta p)A}{r_{\mathrm{c}}\mu(V + V_{\mathrm{eq}})\phi} \tag{4.26}$$

Considering operation at constant pressure drop, which is the most common case, integrating Equation (4.26) gives

$$\frac{t}{V} = \frac{r_{\mathrm{c}}\phi\mu}{2A^2(-\Delta p)}V + \frac{r_{\mathrm{c}}\phi\mu}{A^2(-\Delta p)}V_{\mathrm{eq}} \tag{4.27}$$

## 4.2.4   Washing the Cake

The solid particles separated by filtration often must be washed to remove filtrate from the pores. There are two processes involved in washing. Much of the filtrate occupying the voids between particles may be removed by displacement as clean solvent is passed through the cake. Removal of filtrate held in less accessible regions of the cake and from pores in the particles takes place by diffusion into the wash water. Figure 4.2 shows how the filtrate concentration in the wash solvent leaving the cake varies typically with volume of wash solvent passed.

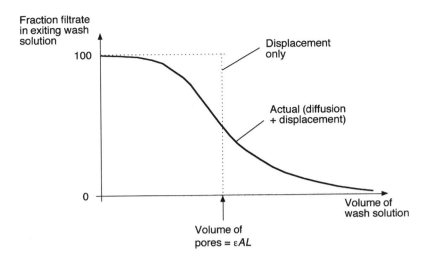

**Figure 4.2**   Removal of filtrate during washing of the filter cake

## 4.2.5   Compressible Cake

In practice many materials give rise to compressible filter cakes. A compressible cake is one whose cake resistance $r_c$ increases with applied pressure difference $(-\Delta p)$. Change in $r_c$ is due mainly to the effect on the cake voidage (recall Equation (4.17)). Fluid drag on the particles in the cake causes a force which is transmitted through the bed. Particles deeper in the bed experience the sum of the forces acting on the particles above. The force on the particles causes the particle packing to become more dense, i.e. cake voidage decreases. In the case of soft particles, the shape or size of the particles may change, adding to the increase in cake resistance.

Referring to Figure 4.3, liquid flows at a superficial velocity $U$ through a

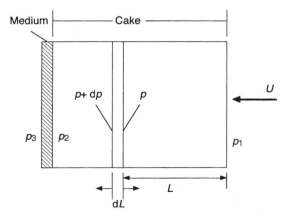

**Figure 4.3**   Analysis of the pressure drop – flow relationship for a compressible cake

filter cake of thickness $H$. Consider an element of the filter cake of thickness $dL$ across which the pressure drop is $dp$. Applying the Carman–Kozeny equation (Equation (4.18)) for flow through this element,

$$-\frac{dp}{dL} = r_c \mu U \tag{4.28}$$

where $r_c$ is the resistance of this element of the cake. For a compressible cake, $r_c$ is a function of the pressure difference between the upstream surface of the cake and the element (i.e., referring to Figure 4.3, $p_1 - p$).

$$\text{Letting} \qquad p_s = p_1 - p \tag{4.29}$$

$$\text{then} \quad -dp = dp_s \tag{4.30}$$

and Equation (4.28) becomes

$$\frac{dp_s}{dL} = r_c \mu U \tag{4.31}$$

In practice the relationship between $r_c$ and $p_s$ must be found from laboratory experiments before Equation (4.29) can be used in design.

## 4.3 FURTHER READING

For further information on fluid flow through packed beds and on filtration the reader is referred to the following:

Coulson, J. M. and Richardson, J. R. (1991), *Chemical Engineering*, Vol. 2, '*Particle Technology and Separation Processes*', 4th Edition, Pergamon, Oxford.
Perry, R. H. and Green, D. (eds) (1984) '*Perry's Chemical Engineering Handbook*', 6th or later editions McGraw-Hill, New York.

## 4.4 WORKED EXAMPLES

### WORKED EXAMPLE 4.1

Water flows through 3.6 kg of glass particles of density 2590 kg/m$^3$ forming a packed bed of depth 0.475 m and diameter 0.0757 m. The variation in frictional pressure drop across the bed with water flowrate in the range 200–1200 cm$^3$/min is shown in columns one and two in Table 4.W1.1

(a)  Demonstrate that the flow is laminar.

(b)  Estimate the mean surface-volume diameter of the particles

(c)  Calculate the relevant Reynolds number.

**Table 4.W1.1**

| Water flowrate (cm³/min) | Pressure drop (mmHg) | $U$ (m/s × 10⁴) | Pressure drop (Pa) |
| --- | --- | --- | --- |
| 200 | 5.5 | 7.41 | 734 |
| 400 | 12.0 | 14.81 | 1600 |
| 500 | 14.5 | 18.52 | 1935 |
| 700 | 20.5 | 25.92 | 2735 |
| 1000 | 29.5 | 37.00 | 3936 |
| 1200 | 36.5 | 44.40 | 4870 |

## Solution

(a) Firstly convert the volumetric water flowrate values into superficial velocities and the pressure drop in mm of mercury into Pascal. These values are shown in columns 3 and 4 of the table.

If the flow is laminar then the pressure gradient across the packed bed should increase linearly with superficial fluid velocity, assuming constant bed voidage and fluid viscosity. Under laminar conditions, the Ergun equation (Equation (4.15)) reduces to

$$\frac{(-\Delta p)}{H} = 150 \frac{\mu U}{x_{sv}^2} \frac{(1 - \varepsilon)^2}{\varepsilon^3}$$

Hence, since the bed depth $H$, the water viscosity $\mu$ and the packed bed voidage $\varepsilon$ may be assumed constant, then $(-\Delta p)$ plotted against $U$ should give a straight line of gradient

$$150 \frac{\mu H}{x_{sv}^2} \frac{(1 - \varepsilon)^2}{\varepsilon^3}$$

This plot is shown in Figure 4.W1.1. The data points fall reasonably on a

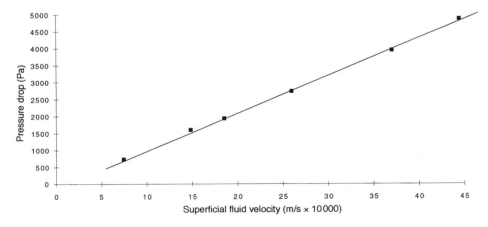

**Figure 4.W1.1**   Plot of packed bed pressure drop versus superficial fluid velocity

straight line confirming laminar flow. The gradient of the straight line is $1.12 \times 10^6$ Pa.s/m and so:

$$150 \frac{\mu H (1 - \varepsilon)^2}{x_{sv}^2 \quad \varepsilon^3} = 1.12 \times 10^6 \text{ Pa.s/m}$$

(b) Knowing the mass of particles in the bed, the density of the particles and the volume of the bed, the voidage may be calculated:

$$\text{mass of bed} = AH[1 - \varepsilon]\rho_p$$

$$\text{giving } \varepsilon = 0.3497$$

Substituting $\varepsilon = 0.3497$, $H = 0.475$ m and $\mu = 0.001$ Pa s in the expression for the gradient of the straight line, we have

$$x_{sv} = 792 \text{ } \mu m$$

(c) The relevant Reynolds number is $Re^* = \dfrac{x U \rho_f}{\mu(1 - \varepsilon)}$ (Equation (4.12))

giving $Re^* = 5.4$ (for the maximum velocity used). This is less than the limiting value for laminar flow (10); a further confirmation of laminar flow.

## WORKED EXAMPLE 4.2

A leaf filter has an area of $0.5$ m$^2$ and operates at a constant pressure drop of 500 kPa. The following test results were obtained for a slurry in water which gave rise to a filter cake regarded as incompressible:

| Volume of filtrate collected (m³) | 0.1 | 0.2 | 0.3 | 0.4 | 0.5 |
|---|---|---|---|---|---|
| Time (s) | 140 | 360 | 660 | 1040 | 1500 |

Calculate:

(a)  the time need to collect $0.8$ m$^3$ of filtrate at a constant pressure drop of 700 kPa.

(b)  the time required to wash the resulting cake with $0.3$ m$^3$ of water at a pressure drop of 400 kPa.

## Solution

For filtration at constant pressure drop we use Equation (4.27), which indicates that if we plot $t/V$ versus $V$ a straight line will have a gradient

$$\frac{r_c \phi \mu}{2A^2(-\Delta p)}$$

and an intercept $\dfrac{r_c \phi \mu}{A^2(-\Delta p)} V_{eq}$ on the $t/V$ axis.

Using the data given in the question:

| V (m³) | 0.1 | 0.2 | 0.3 | 0.4 | 0.5 |
|---|---|---|---|---|---|
| t/V (s/m³) | 1400 | 1800 | 2200 | 2600 | 3000 |

This is plotted in Figure 4.W2.1.

From the plot:      gradient $= 4000 \text{ s/m}^6$

intercept $= 1000 \text{ s/m}^3$

$$\text{hence, } \frac{r_c \phi \mu}{2 A^2 (-\Delta p)} = 4000$$

$$\text{and } \frac{r_c \phi \mu}{A^2 (-\Delta p)} V_{eq} = 1000$$

which, with $A = 0.5 \text{ m}^2$ and $(-\Delta p) = 500 \times 10^3$ Pa, gives

$$r_c \phi \mu = 1 \times 10^9 \text{ Pa s/m}^2$$

and $V_{eq} = 0.125 \text{ m}^3$
  Substituting in Equation (4.27),

$$\frac{t}{V} = \frac{0.5 \times 10^9}{(-\Delta p)} (4V + 1)$$

which applies to the filtration of the same slurry in the same filter at any pressure drop.

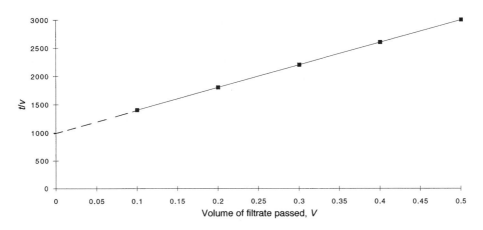

**Figure 4.W2.1**    Plot of t/V versus V for worked example 4.2

(a) To calculate the time required to pass $0.8 \, \text{m}^3$ of filtrate at a pressure drop of 700 kPa, we substitute $V = 0.8 \, \text{m}^3$ and $(-\Delta p) = 700 \times 10^3$ Pa in the above equation, giving

$$t = 2400 \, \text{s (or 40 min)}$$

(b) During the filtration the cake thickness is continuously increasing and, since the pressure drop is constant, the volume flowrate of filtrate will continuously decrease. The filtration rate is given by Equation (4.26). Substituting the volume of filtrate passed at the end of the filtration period ($V = 0.8 \, \text{m}^3$), $r_c \phi \mu = 1 \times 10^9 \, \text{Pa s/m}^2$, $V_{eq} = 0.125 \, \text{m}^3$ and $(-\Delta p) = 700 \times 10^3$ Pa, we find the filtration rate at the end of the filtration period is $dV/dt = 1.89 \times 10^{-4} \, \text{m}^3/\text{s}$

If we assume that the wash water has the same physical properties as the filtrate, then during a wash period at a pressure drop of 700 kPa the wash rate is also $1.89 \times 10^{-4} \, \text{m}^3/\text{s}$. However, the applied pressure drop during the wash cycle is 400 kPa. According to Equation (4.26) the liquid flowrate is directly proportional to the applied pressure drop, and so

$$\text{flowrate of wash water (at 400 kPa)} = 1.89 \times 10^{-4} \times \left( \frac{400 \times 10^3}{700 \times 10^3} \right)$$

$$= 1.08 \times 10^{-4} \, \text{m}^3/\text{s}$$

Hence, the time needed to pass $0.3 \, \text{m}^3$ of wash water at this rate is 2778 s (or 46.3 min)

## EXERCISES

**4.1** A packed bed of solid particles of density $2500 \, \text{kg/m}^3$ occupies a depth of 1 m in a vessel of cross-sectional area $0.04 \, \text{m}^2$. The mass of solids in the bed is 50 kg and the surface-volume mean diameter of the particles is 1 mm. A liquid of density $800 \, \text{kg/m}^3$ and viscosity 0.002 Pa s flows upwards through the bed, which is restrained at its upper surface.

(a)  Calculate the voidage (volume fraction occupied by voids) of the bed.

(b)  Calculate the pressure drop across the bed when the volume flow rate of liquid is $1.44 \, \text{m}^3/\text{h}$. (Answer: (a) 0.50; (b) 6560 Pa [Ergun])

**4.2** A packed bed of solids of density $2000 \, \text{kg/m}^3$ occupies a depth of 0.6 m in a cylindrical vessel of inside diameter 0.1 m. The mass of solids in the bed is 5 kg and the surface-volume mean diameter of the particles is 300 μm. Water (density $1000 \, \text{kg/m}^3$ and viscosity 0.001 Pa s) flows upwards through the bed.

(a)  What is the voidage of the packed bed?

(b)  Calculate the superficial liquid velocity at which the pressure drop across the bed is 4130 Pa. (Answer: (a) 0.4692; (b) 1.5 mm/s, using Ergun)

**4.3** A gas absorption tower of diameter 2 m contains ceramic Raschig rings randomly packed to a height of 5 m. Air containing a small proportion of sulphur dioxide passes upwards through the absorption tower at a flow rate of 6 m$^3$/s. The viscosity and density of the gas may be taken as $1.80 \times 10^{-5}$ Pa s and 1.2 kg/m$^3$ respectively. Details of the packing are given below:

Ceramic Raschig rings

    surface area per unit volume of packed bed, $S_B = 190$ m$^2$/m$^3$;

    voidage of randomly packed bed $= 0.71$

(a)   Calculate the diameter, $d_{sv}$, of a sphere with the same surface-volume ratio as the Raschig rings.

(b)   Calculate the pressure drop across the packing in the tower.

(c)   *Discuss* how this pressure drop will vary with flow rate of the gas within $\pm 10\%$ of the quoted flow rate.

(d)   *Discuss* how the pressure drop across the packing would vary with gas pressure and temperature. (Answer: (a) 9.16 mm; (b) 3460 Pa; for (c), (d) use the hint that turbulence dominates)

**4.4** A solution of density 1100 kg/m$^3$ and viscosity $2 \times 10^{-3}$ Pa s is flowing under gravity at a rate of 0.24 kg/s through a bed of catalyst particles. The bed diameter is 0.2 m and the depth is 0.5 m. The particles are cylindrical, with a diameter of 1 mm and length of 2 mm. They are loosely packed to give a voidage of 0.3. Calculate the depth of liquid above the top of the bed. (*Hint*: apply the mechanical energy equation between the bottom of the bed and the surface of the liquid.)

(Answer: height of liquid above bed $= 0.716$ m)

**4.5** In the regeneration of an ion exchange resin, hydrochloric acid of density 1200 kg/m$^3$ and viscosity $2 \times 10^{-3}$ Pa s flows upwards through a bed of resin particles of density 2500 kg/m$^3$ resting on a porous support in a tube 4 cm in diameter. The particles are spherical, have a diameter 0.2 mm and form a bed of void fraction 0.5. The bed is 60 cm deep and is unrestrained at its upper surface. Plot the pressure drop across the bed as function of acid flowrate up to a value of 0.1 l/min.

(Answer: Pressure drop increases linearly up to a value of 3826 Pa beyond which point the bed will fluidize and maintain this pressure drop (see Chapter 5))

**4.6** The reactor of a catalytic reformer contains spherical catalyst particles of diameter 1.46 mm. The packed volume of the reactor is to be 3.4 m$^3$ and the void fraction is 0.25. The reactor feed is a gas of density 30 kg/m$^3$ and viscosity

$2 \times 10^{-5}$ Pa s flowing at a rate of 11 320 m$^3$/h. The gas properties may be assumed constant. The pressure loss through the reactor is restricted to 68.95 kPa. Calculate the cross-sectional area for flow and the bed depth required.

(Answer: area $= 9.71$ m$^2$; depth $= 0.35$ m)

**4.7** A leaf filter has an area of 2 m$^2$ and operates at a constant pressure drop of 250 kPa. The following results were obtained during a test with an incompressible cake:

| Volume of filtrate collected (litres) | 280 | 430 | 540 | 680 | 800 |
|---|---|---|---|---|---|
| Time (min) | 10 | 20 | 30 | 45 | 60 |

Calculate:

(a) the time required to collect 1200 litre of filtrate at a constant pressure drop of 400 kPa with the same feed slurry.

(b) the time required to wash the resulting filter cake with 500 litre of water (same properties as the filtrate) at a pressure drop of 200 kPa.

(Answer: (a) 79.4 min; (b) 124 min)

**4.8** A laboratory leaf filter has an area of 0.1 m$^2$, operates at a constant pressure drop of 400 kPa and produces the following results during a test on filtration of a slurry:

| Volume of filtrate collected (litres) | 19 | 31 | 41 | 49 | 56 | 63 |
|---|---|---|---|---|---|---|
| Time (s) | 300 | 600 | 900 | 1200 | 1500 | 1800 |

(a) Calculate the time required to collect 1.5 m$^3$ of filtrate during filtration of the same slurry at a constant pressure drop of 300 kPa on a similar full-scale filter with an area of 2 m$^2$.

(b) Calculate the rate of passage of filtrate at the end of the filtration in part (a).

(c) Calculate the time required to wash the resulting filter cake with 0.5 m$^3$ of water at a constant pressure drop of 200 kPa.

(Assume the cake is incompressible and that the flow properties of the filtrate are the same as those of the wash solution.)

(Answer: (a) 37.2 min; (b) 20.4 litre/min; (c) 36.7 min)

**4.9** A leaf filter has an area of 1.73 m$^2$, operates at a constant pressure drop of 300 kPa and produces the following results during a test on filtration of a slurry:

| Volume of filtrate collected (m³) | 0.19 | 0.31 | 0.41 | 0.49 | 0.56 | 0.63 |
|---|---|---|---|---|---|---|
| Time (s) | 300 | 600 | 900 | 1200 | 1500 | 1800 |

Assuming that the cake is incompressible and that the flow properties of the filtrate are the same as those of the wash solution, calculate:

(a)   the time required to collect 1 m³ of filtrate during filtration of the same slurry at a constant pressure drop of 400 kPa.

(c)   the time required to wash the resulting filter cake with 0.8 m³ of water at a constant pressure drop of 250 kPa.

(Answer: (a) 49.5 min; (b) 110.9 min)

# 5

# Fluidization

## 5.1 FUNDAMENTALS

When a fluid is passed upwards through a bed of particles the pressure loss in the fluid due to frictional resistance increases with increasing fluid flow. A point is reached when the upward drag force exerted by the fluid on the particles is equal to the apparent weight of particles in the bed. At this point the particles are lifted by the fluid, the separation of the particles increases, and the bed becomes fluidized. The force balance across the fluidized bed dictates that the fluid pressure loss across the bed of particles is equal to the apparent weight of the particles per unit area of the bed. Thus:

$$\text{pressure drop} = \frac{\text{weight of particles} - \text{upthrust on particles}}{\text{bed cross-sectional area}}$$

For a bed of particles of density $\rho_p$, fluidized by a fluid of density $\rho_f$ to form a bed of depth $H$ and voidage $\varepsilon$ in a vessel of cross-sectional area $A$:

$$\Delta p = \frac{HA(1 - \varepsilon)(\rho_p - \rho_f)g}{A} \tag{5.1}$$

or,

$$\Delta p = H(1 - \varepsilon)(\rho_p - \rho_f)g \tag{5.2}$$

A plot of fluid pressure loss across the bed versus superficial fluid velocity through the bed would have the appearance of Figure 5.1. Referring to this figure, the straight line region OA is the packed bed region. Here the solid particles do not move relative to one another and their separation is constant. The pressure loss versus fluid velocity relationship in this region is described by the Carman–Kozeny equation (Equation (4.9)) in the laminar flow regime and the Ergun equation in general (Equation 4.11). (See Chapter 4 for a detailed analysis of packed bed flow).

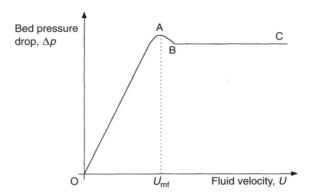

**Figure 5.1**   Pressure drop versus fluid velocity for packed and fluidized beds

The region BC is the fluidized bed region where Equation (5.1) applies. At point A it will be noticed that the pressure loss rises above the value predicted by Equation (5.1). This rise is more marked in powders which have been compacted to some extent before the test and is associated with the extra force required to overcome interparticle attractive forces.

The superficial fluid velocity at which the packed bed becomes a fluidized bed is known as the minimum fluidization velocity, $U_{mf}$. This is also some-times referred to as the velocity at incipient fluidization (incipient meaning beginning). $U_{mf}$ increases with particle size and particle density and is affected by fluid properties. It is possible to derive an expression for $U_{mf}$ by equating the expression for pressure loss in a fluidized bed (Equation (5.2)) with the expression for pressure loss across a packed bed. Thus recalling the Ergun equation (Equation (4.11)):

$$\frac{(-\Delta p)}{H} = 150\frac{(1-\varepsilon)^2}{\varepsilon^3}\frac{\mu U}{x_{sv}^2} + 1.75\frac{(1-\varepsilon)}{\varepsilon^3}\frac{\rho_f U^2}{x_{sv}} \qquad (5.3)$$

substituting the expression for $(-\Delta p)$ from Equation (5.2):

$$(1-\varepsilon)(\rho_p - \rho_f)g = 150\frac{(1-\varepsilon)^2}{\varepsilon^3}\frac{\mu U_{mf}}{x_{sv}^2} + 1.75\frac{(1-\varepsilon)}{\varepsilon^3}\frac{\rho_f U_{mf}^2}{x_{sv}} \qquad (5.4)$$

Rearranging,

$$(1-\varepsilon)(\rho_p - \rho_f)g = 150\frac{(1-\varepsilon)^2}{\varepsilon^3}\left(\frac{\mu^2}{\rho_f x_{sv}^3}\right)\left(\frac{U_{mf} x_{sv} \rho_f}{\mu}\right)$$

$$+ 1.75\frac{(1-\varepsilon)}{\varepsilon^3}\left(\frac{\mu^2}{\rho_f x_{sv}^3}\right)\left(\frac{U_{mf}^2 x_{sv}^2 \rho_f^2}{\mu^2}\right) \qquad (5.5)$$

and so,

$$(1 - \varepsilon)(\rho_p - \rho_f)g\left(\frac{\rho_f x_{sv}^3}{\mu^2}\right) = 150\frac{(1 - \varepsilon)^2}{\varepsilon^3} Re_{mf} + 1.75\frac{(1 - \varepsilon)}{\varepsilon^3} Re_{mf}^2 \qquad (5.6)$$

or,

$$Ar = 150\frac{(1 - \varepsilon)}{\varepsilon^3} Re_{mf} + 1.75\frac{1}{\varepsilon^3} Re_{mf}^2 \qquad (5.7)$$

where $Ar$ is the dimensionless number known as the Archimedes number,

$$Ar = \frac{\rho_f(\rho_p - \rho_f)gx_{sv}^3}{\mu^2}$$

and $Re_{mf}$ is the Reynolds number at incipient fluidization,

$$Re_{mf} = \left(\frac{U_{mf}x_{sv}\rho_f}{\mu}\right)$$

In order to obtain a value of $U_{mf}$ from Equation (5.7) we need to know the voidage of the bed at incipient fluidization, $\varepsilon = \varepsilon_{mf}$. Taking $\varepsilon_{mf}$ as the voidage of the packed bed, we can obtain a crude $U_{mf}$. However, in practice voidage at the onset of fluidization may be considerably greater than the packed bed voidage. A typical often used value of $\varepsilon_{mf}$ is 0.4. Using this value, Equation (5.7) becomes

$$Ar = 1652 \ Re_{mf} + 24.51 \ Re_{mf}^2 \qquad (5.8)$$

Wen and Yu (1966) produced an empirical correlation for $U_{mf}$ with a form similar to Equation 5.8:

$$Ar = 1652Re_{mf} + 24.51Re_{mf}^2 \qquad (5.9)$$

The Wen and Yu correlation is often expressed in the form:

$$Re_{mf} = 33.7[(1 + 3.59 \times 10^{-5}Ar)^{0.5} - 1] \qquad (5.10)$$

and is valid for spheres in the range $0.01 < Re_{mf} < 1000$.

For gas fluidization the Wen and Yu correlation is often taken as being most suitable for particles larger than 100 μm, whereas the correlation of Baeyens and Geldart (1974), shown below in Equation (5.11), is best for particles less than 100 μm.

$$U_{mf} = \frac{(\rho_p - \rho_f)^{0.934} g^{0.934} x_p^{1.8}}{1110\mu^{0.87}\rho_g^{0.066}} \qquad (5.11)$$

## 5.2   RELEVANT POWDER AND PARTICLE PROPERTIES

The correct density for use in fluidization equations is the particle density, defined as the mass of a particle divided by its hydrodynamic volume. This is the volume 'seen' by the fluid in its fluid dynamic interaction with the particle and includes the volume of all the open and closed pores (see Figure 5.2):

$$\text{particle density} = \frac{\text{mass of particle}}{\text{hydrodynamic volume of particle}}$$

For non-porous solids, this is easily measured by a gas pycnometer or specific gravity bottle, but these devices should not be used for porous solids since they give the true or absolute density $\rho_{abs}$ of the material of which the particle is made and this is not appropriate where interaction with fluid flow is concerned:

$$\text{absolute density} = \frac{\text{mass of particle}}{\text{volume of solids material making up the particle}}$$

For porous particles, the particle density $\rho_p$ (also called apparent or envelope density) is not easy to measure directly although several methods are given in Geldart (1990). Bed density is another term used in connection with fluidized beds; bed density is defined as

$$\text{bed density} = \frac{\text{mass of particles in a bed}}{\text{volume occupied by particles and voids between them}}$$

For example, 600 kg of powder is fluidized in a vessel of cross-sectional area 1 m² and achieves a bed height of 0.5 m. What is the bed density?

Mass of particles in the bed = 600 kg.

Volume occupied by particles and voids = $1 \times 0.5 = 0.5$ m³.

Hence, bed density = $600/0.5 = 1200$ kg/m³
If the particle density of these solids is 2700 kg/m³, what is the bed voidage?

Bed density $\rho_B$ is related to particle density $\rho_p$ and bed voidage $\varepsilon$ by Equation (5.12):

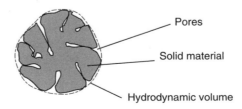

**Figure 5.2**   Hydrodynamic volume of a particle

$$\rho_B = (1 - \varepsilon)\rho_p \tag{5.12}$$

Hence, voidage $= 1 - \frac{1200}{2700} = 0.555$.

Another density often used when dealing with powders is the bulk density: it is defined in a similar way to fluid bed density:

$$\text{bulk density} = \frac{\text{mass of particles}}{\text{volume occupied by particles and voids between them}}$$

The most appropriate particle size to use in equations relating to fluid–particle interactions is a hydrodynamic diameter; i.e. an equivalent sphere diameter derived from a measurement technique involving hydrodynamic interaction between the particle and fluid. In practice, however, in most industrial applications sizing is done using sieving and correlations use either sieve diameter, $x_p$ or volume diameter, $x_v$. For spherical or near spherical particles $x_v$ is equal to $x_p$. For angular particles, $x_v \approx 1.13x_p$.

For use in fluidization applications, starting from a sieve analysis the mean size of the powder is often calculated from

$$\text{mean } x_p = \frac{1}{\sum m_i/x_i} \tag{5.13}$$

where $x_i$ is the arithmetic mean of adjacent sieves between which a mass fraction $m_i$ is collected. This is the harmonic mean of the mass distribution, which was shown in Chapter 3 to be equivalent to the arithmetic mean of a surface distribution.

## 5.3  BUBBLING AND NON-BUBBLING FLUIDIZATION

Beyond the minimum fluidization velocity bubbles or particle-free voids may appear in the fluidized bed. Figure 5.3 shows bubbles in a gas fluidized bed. The equipment used in the figure is a so-called 'two-dimensional fluidized bed'. A favourite tool of researchers looking at bubble behaviour, this is actually a vessel of a rectangular cross-section, whose shortest dimension (into the page) is usually only 1 cm or so.

At superficial velocities above the minimum fluidization velocity, fluidization may in general be either bubbling or non-bubbling. Some combinations of fluid and particles give rise to *only bubbling* fluidization and some combinations give *only non-bubbling* fluidization. Most liquid fluidized systems, except those involving very dense particles, do not give rise to bubbling. Figure 5.4 shows a bed of glass spheres fluidized by water exhibiting non-bubbling fluidized bed behaviour. Gas fluidized systems, however, give either only bubbling fluidization or non-bubbling fluidization beginning at $U_{mf}$, followed by bubbling fluidization as fluidizing velocity increases. Non-bubbling fluidization is also known as particulate or homogeneous fluidization and bubbling fluidization is often referred to as aggregative or heterogeneous fluidization.

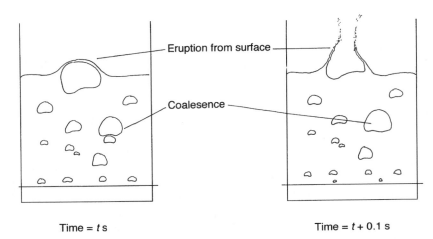

Time = $t$ s                                                        Time = $t + 0.1$ s

**Figure 5.3**  Sequence showing bubbles in a 'two-dimensional' fluidized bed of Group B powder. Sketches taken from video

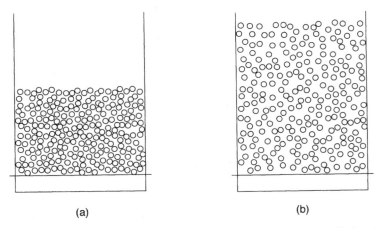

(a)                                                                                    (b)

**Figure 5.4**  Expansion of a liquid fluidized bed: (a) just above $U_{mf}$ (b) liquid velocity several times $U_{mf}$. Note uniform increase in void fraciton. Sketches taken from video

## 5.4  CLASSIFICATION OF POWDERS

Geldart (1973) classified powders into four groups according to their fluidization properties at ambient conditions. The Geldart classification of powders is
now used widely in all fields of powder technology. Powders which when
fluidized by air at ambient conditions give a region of non-bubbling fluidization beginning at $U_{mf}$, followed by bubbling fluidization as fluidizing velocity
increases, are classified as Group A. Powders which under these conditions
give only bubbling fluidization are classified as Group B. Geldart identified
two further groups; Group C powders – very fine, cohesive powders which
are incapable of fluidization in the strict sense, and Group D powders – large

particles distinguished by their ability to produce deep spouting beds (see Figure 5.5). Figure 5.6 shows how the group classifications are related to the particle and gas properties.

The fluidization properties of a powder in air may be predicted by establishing in which group it lies. It is important to note that at operating temperatures and pressures above ambient a powder may appear in a different group from that which it occupies at ambient conditions. This is due to the effect of gas properties on the grouping and may have serious implications as far as the operation of the fluidized bed is concerned. Table 5.1 presents a summary of the typical properties of the different powder classes.

Since the range of gas velocities over which non-bubbling fluidization occurs in Group A powders is small, bubbling fluidization is the type most commonly encountered in gas fluidized systems in commercial use. The superficial gas velocity at which bubbles first appear is known as the minimum bubbling velocity $U_{mb}$. Premature bubbling can be caused by poor distributor design or protuberances inside the bed. Abrahamsen and Geldart (1980) correlated the maximum values of $U_{mb}$ with gas and particle properties using the following correlation:

**Figure 5.5**   A spouted fluidized bed of rice

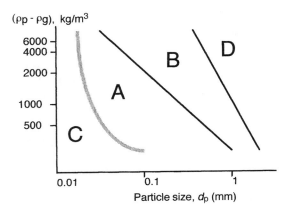

**Figure 5.6**   Geldart's classification of powders

$$U_{mb} = 2.07 \exp(0.716F)\left[\frac{x_p \rho_g^{0.06}}{\mu^{0.347}}\right]$$   (5.14)

where $F$ is the fraction of powder less than 45 μm.

In Group A powders $U_{mb} > U_{mf}$, bubbles are constantly splitting and coalescing, and a maximum stable bubble size is achieved. This makes for good quality, smooth fluidization. Figure 5.7 shows bubbles in a Group A powder in a two-dimensional fluidized bed.

In Groups B and D powders $U_{mb} = U_{mf}$, bubbles continue to grow, never achieving a maximum size (see Figure 5.3). This makes for rather poor quality fluidization associated with large pressure fluctuations.

In Group C powders the interparticle forces are large compared with the inertial forces on the particles. As a result, the particles are unable to achieve the separation they require to be totally supported by drag and buoyancy forces and true fluidization does not occur. Bubbles, as such, do not appear; instead the gas flow forms channels through the powder (see Figure 5.8). Since the particles are not fully supported by the gas, the pressure loss across the bed is always less than apparent weight of the bed per unit cross-sectional area. Consequently, measurement of bed pressure drop is one means of detecting this Group C behaviour if visual observation is inconclusive. Fluidization, of sorts, can be achieved with the assistance of a mechanical stirrer or vibration.

When the size of the bubbles is greater than about one third of the diameter of the equipment their rise velocity is controlled by the equipment and they become slugs of gas. Slugging is attended by large pressure fluctuations and so it is generally avoided in large units since it can cause vibration to the plant. Slugging is unlikely to occur at any velocity if the bed is sufficiently shallow. According to Yagi and Muchi (1952), slugging will not occur provided the following criterion is satisfied:

$$\left[\frac{H_{mf}}{D}\right] \leqslant \frac{1.9}{(\rho_p x_p)^{0.3}}$$   (5.15)

**Table 5.1** Geldart's classification of powders

| | Group C | Group A | Group B | Group D |
|---|---|---|---|---|
| Most obvious characteristic | Cohesive, difficult to fluidize | Ideal for fluidization. Exhibits range of non-bubbling fluidization | Starts bubbling at $U_{mf}$ | Coarse solids |
| Typical solids | Flour, cement | Cracking catalyst | Building sand | Gravel, coffee beans |
| *Property* | | | | |
| Bed expansion | Low because of chanelling | High | Moderate | Low |
| De-aeration rate | Initially fast, then exponential | Slow, linear | Fast | Fast |
| Bubble properties | No bubbles–only channels | Bubbles split and coalesce. Maximum bubble size. | No limit to size | No limit to size |
| Solids mixing | Very low | High | Moderate | Low |
| Gas backmixing | Very low | High | Moderate | Low |
| Spouting | No | No | Only in shallow beds | Yes, even in deep beds |

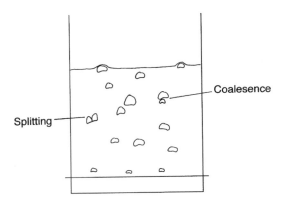

**Figure 5.7** Bubbles in a 'two-dimensional' fluidized bed of Group A powder. Sketch taken from video

**Figure 5.8** Attempts to fluidize Group C powder producing cracks and channels or discrete solid plugs!

This criterion works well for most powders. If the bed is deeper than this critical height then slugging will occur when the gas velocity exceeds $U_{ms}$ as given by Equation (5.16) (Baeyens and Geldart, 1974):

$$U_{ms} = U_{mf} + 0.16(1.34D^{0.175} - H_{mf})^2 + 0.07(gD)^{0.5} \qquad (5.16)$$

## 5.5 EXPANSION OF A FLUIDIZED BED

### 5.5.1 Non-bubbling Fluidization

In a non-bubbling fluidized bed beyond $U_{mf}$ the particle separation increases with increasing fluid superficial velocity whilst the pressure loss across the bed remains constant. This increase in bed voidage with fluidizing velocity is referred to as bed expansion (see Figure 5.4). The relationship between fluid velocity and bed voidage may be determined by recalling the analysis of multiple particle systems (Chapter 2). For a particle suspension settling in a fluid under force balance conditions the relative velocity $U_{rel}$ between particles and fluid is given by

$$U_{rel} = U_p - U_f = U_T \varepsilon f(\varepsilon) \tag{5.17}$$

where $U_p$ and $U_f$ are the actual downward vertical velocities of the particles and the fluid, and $U_T$ is the single particle terminal velocity in the fluid. In the case of a fluidized bed the time-averaged actual vertical particle velocity is zero ($U_p = 0$) and so

$$U_f = -U_T \varepsilon f(\varepsilon) \tag{5.18}$$

or

$$U_{fs} = -U_T \varepsilon^2 f(\varepsilon) \tag{5.19}$$

where $U_{fs}$ is the downward volumetric fluid flux. In common with fluidization practice, we will use the term superficial velocity ($U$) rather than volumetric fluid flux. Since the upward superficial fluid velocity (U) is equal to the upward volumetric fluid flux ($-U_{fs}$), and $U_{fs} = U_f \varepsilon$, then:

$$U = U_T \varepsilon^2 f(\varepsilon) \tag{5.20}$$

Richardson and Zaki (1954) found the function $f(\varepsilon)$ which applied to both hindered settling and to non-bubbling fluidization. They found that in general, $f(\varepsilon) = \varepsilon^n$, where the exponent $n$ was independent of particle Reynolds number at very low Reynolds numbers, when the drag force is independent of fluid density, and at high Reynolds number, when the drag force is independent of fluid viscosity, i.e.

$$\text{in general: } U = U_T \varepsilon^n \tag{5.21}$$

$$\text{For } Re_p \leqslant 0.3; \; f(\varepsilon) = \varepsilon^{2.65} \Rightarrow U = U_T \varepsilon^{4.65} \tag{5.22}$$

$$\text{For } Re_p \geqslant 500; \; f(\varepsilon) = \varepsilon^{0.4} \Rightarrow U = U_T \varepsilon^{2.4} \tag{5.23}$$

Richardson and Khan (1989) suggested the correlation given in Equation (2.25) (Chapter 2) which permits the determination of the exponent $n$ at intermediate values of Reynolds number (although it is expressed in terms of the

Archimedes number $Ar$ there is a direct relationship between $Re_p$ and $Ar$). This correlation also incorporates the effect of the vessel diameter on the exponent. Thus Equations (5.21), (5.22) and (5.23) in conjunction with Equation (2.25) permit calculation of the variation in bed voidage with fluid velocity beyond $U_{mf}$. Knowledge of the bed voidage allows calculation of the fluidized bed height as illustrated below:

$$\text{mass of particles in the bed} = M_B = (1 - \varepsilon)\rho_p AH \qquad (5.24)$$

If packed bed depth ($H_1$) and voidage ($\varepsilon_1$) are known, then if the mass remains constant the bed depth at any voidage can be determined:

$$(1 - \varepsilon_2)\rho_p AH_2 = (1 - \varepsilon_1)\rho_p AH_1 \qquad (5.25)$$

hence,

$$H_2 = \frac{(1 - \varepsilon_1)}{(1 - \varepsilon_2)} H_1$$

## 5.5.2  Bubbling Fluidization

The simplest description of the expansion of a bubbling fluidized bed is derived from the two-phase theory of fluidization of Toomey and Johnstone (1952). This theory considers the bubbling fluidized bed to be composed of two phases; the bubbling phase (the gas bubbles) and the particulate phase (the fluidized solids around the bubbles). The particulate phase is also referred to as the emulsion phase. The theory states that any gas in excess of that required at incipient fluidization will pass through the bed as bubbles. Figure 5.9 shows the effect of fluidizing gas velocity on bed expansion of a Group A

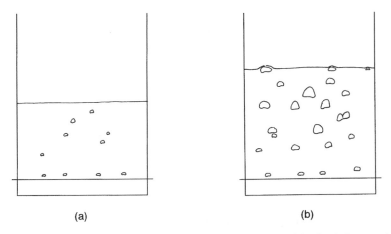

(a)                                                          (b)

**Figure 5.9**  Bed expansion in a 'two-dimensional' fluidized bed of Group A powder: (a) just above $U_{mb}$; (b) fluidized at several times $U_{mb}$. Sketches taken from video

powder fluidized by air. Thus, referring to Figure 5.10, $Q$ is the actual gas flow rate to the fluid bed and $Q_{mf}$ is the gas flow rate at incipient fluidization, then

$$\text{gas passing through the bed as bubbles} = Q - Q_{mf} = (U - U_{mf})A \quad (5.26)$$

$$\text{gas passing through the emulsion phase} = Q_{mf} = U_{mf}A \quad (5.27)$$

Expressing the bed expansion in terms of the fraction of the bed occupied by bubbles, $\varepsilon_B$:

$$\varepsilon_B = \frac{H - H_{mf}}{H} = \frac{Q - Q_{mf}}{AU_B} = \frac{(U - U_{mf})}{U_B} \quad (5.28)$$

where $H$ is the bed height at $U$ and $H_{mf}$ is the bed height at $U_{mf}$ and $U_B$ is the mean rise velocity of a bubble in the bed (obtained from correlations; see below). The voidage of the emulsion phase is taken to be that at minimum fluidization $\varepsilon_{mf}$. The mean bed voidage is then given by

$$(1 - \varepsilon) = (1 - \varepsilon_B)(1 - \varepsilon_{mf}) \quad (5.29)$$

In practice, the elegant two-phase theory overestimates the volume of gas passing through the bed as bubbles (the visible bubble flow rate) and better estimates of bed expansion may be obtained by replacing $(Q - Q_{mf})$ in Equation (5.28) with

$$\text{visible bubble flow rate, } Q_B = Y(U - U_{mf}) \quad (5.30)$$

$$\text{where } 0.8 < Y < 1.0 \quad \text{For Group A powders}$$

$$0.6 < Y < 0.8 \quad \text{For Group B powders}$$

$$0.25 < Y < 0.6 \quad \text{For Group D powders}$$

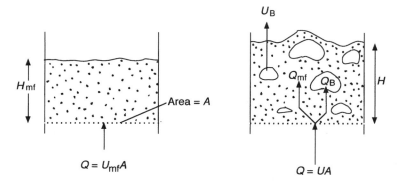

**Figure 5.10**   Gas flows in a fluidized bed according to the two-phase theory

The above analysis requires a knowledge of the bubble rise velocity $U_B$, which depends on the bubble size $d_{Bv}$ and bed diameter $D$. The bubble diameter at a given height above the distributor depends on the orifice density in the distributor $N$, the distance above the distributor $L$ and the excess gas velocity $(U - U_{mf})$.

*For Group B powders*:

$$d_{Bv} = \frac{0.54}{g^{0.2}}(U - U_{mf})^{0.4}(L + 4N^{-0.5})^{0.8} \qquad \{\text{Darton } et\ al., 1977\} \qquad (5.31)$$

$$U_B = \Phi_B(gd_{Bv})^{0.5} \qquad\qquad \{\text{Werther, 1983}\} \qquad (5.32)$$

where,

$$\left\{ \begin{array}{ll} \Phi_B = 0.64 & \text{for } D \leqslant 0.1 \text{ m} \\ \Phi_B = 1.6D^{0.4} & \text{for } 0.1 < D \leqslant 1 \text{ m} \\ \Phi_B = 1.6 & \text{for } D > 1 \text{ m} \end{array} \right\} \qquad (5.33)$$

*For Group A powders*: bubbles reach a maximum stable size which may be estimated from

$$d_{Bv\,max} = 2(U_{T2.7})^2/g \qquad \{\text{Geldart (1992)}\} \qquad (5.34)$$

where $U_{T2.7}$ is the terminal free fall velocity for particles of diameter 2.7 times the actual mean particle diameter.

Bubble velocity for Group A powders is again given by Werther (1983):

$$U_B = \Phi_A(gd_{Bv})^{0.5} \qquad \{\text{Werther (1983)}\} \qquad (5.35)$$

where,

$$\left\{ \begin{array}{ll} \Phi_A = 1 & \text{for } D \leqslant 0.1 \text{ m} \\ \Phi_A = 2.5D^{0.4} & \text{for } 0.1 < D \leqslant 1 \text{ m} \\ \Phi_A = 2.5 & \text{for } D > 1 \text{ m} \end{array} \right\} \qquad (5.36)$$

## 5.6  ENTRAINMENT

The term entrainment will be used here to describe the ejection of particles from the surface of a bubbling bed and their removal from the vessel in the fluidizing gas. In the literature on the subject other terms such as 'carryover' and 'elutriation' are often used to describe the same process. In this section we will study the factors affecting the rate of entrainment of solids from a fluidized bed and develop a simple approach to the estimation of the entrainment rate and the size distribution of entrained solids.

Consider a single particle falling under gravity in a static gas in the absence of any solids boundaries. We know that this particle will reach a terminal velocity when the forces of gravity, buoyancy and drag are balanced (see Chapter 1). If the gas of infinite extent is now considered to be moving upwards at a velocity equal to the terminal velocity of the particle, the particle will be stationary. If the gas is moving upwards in a pipe at a superficial velocity equal to the particle's terminal velocity, then:

(a)  in laminar flow: the particle may move up or down depending on its radial position because of the parabolic velocity profile of the gas in the pipe.

(b)  in turbulent flow: the particle may move up or down depending on its radial position. In addition the random velocity fluctuations superimposed on the time-averaged velocity profile make the actual particle motion less predictable.

If we now introduce into the moving gas stream a number of particles with a range of particle size some particles may fall and some may rise depending on their size and their radial position. Thus the entrainment of particles in an upward-flowing gas stream is a complex process. We can see that the rate of entrainment and the size distribution of entrained particles will in general depend on particle size and density, gas properties, gas velocity, gas flow regime–radial velocity profile and fluctuations and vessel diameter. In addition (i) the mechanisms by which the particles are ejected into the gas stream from the fluidized bed are dependent on the characteristics of the bed – in particular bubble size and velocity at the surface, (ii) the gas velocity profile immediately above the bed surface is distorted by the bursting bubbles. It is not surprising then that prediction of entrainment from first principles is not possible and in practice an empirical approach must be adopted.

This empirical approach defines coarse particles as particles whose terminal velocity is greater than the superficial gas velocity ($U_T > U$) and fine particles as those for which $U_T < U$, and considers the region above the fluidized bed surface to be composed of several zones shown in Figure 5.11:

• *Freeboard*. Region between the bed surface and the gas outlet.

• *Splash zone*. Region just above the bed surface in which coarse particles fall back down.

• *Disengagement zone*. Region above the splash zone in which the upward flux and suspension concentration of fine particles decreases with increasing height.

• *Dilute-phase transport zone*. Region above the disengagement zone in which all particles are carried upwards; particle flux and suspension concentration are constant with height.

Note that, although in general fine particles will be entrained and leave the system and coarse particles will remain, in practice fine particles may stay in the system at velocities several times their terminal velocity and coarse particles may be entrained.

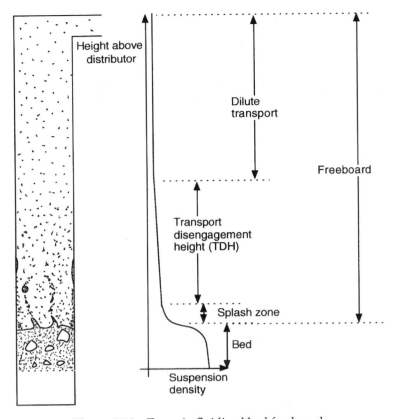

**Figure 5.11**   Zones in fluidized bed freeboard

The height from the bed surface to the top of the disengagement zone is known as the transport disengagement height (TDH). Above TDH the entrainment flux and concentration of particles is constant. Thus, from the design point of view, in order to gain maximum benefit from the effect of gravity in the freeboard, the gas exit should be placed above the TDH. Many empirical correlations for TDH are available in the literature; those of Horio *et al.* (1980) presented in Equation (5.37) and Zenz (1983) presented graphically in Figure 5.12 are two of the more reliable ones.

$$\text{TDH} = 4.47 d_{\text{Bvs}}^{0.5} \tag{5.37}$$

($d_{\text{bvs}}$ = equivalent volume diameter of a bubble at the surface).

The empirical estimation of entrainment rates from fluidized beds is based on the following rather intuitive equation:

$$\begin{bmatrix} \text{instantaneous rate of loss} \\ \text{of solids of size } x_i \end{bmatrix} \propto \text{bed area} \times \begin{bmatrix} \text{fraction of bed with} \\ \text{size } x_i \text{ at time } t \end{bmatrix}$$

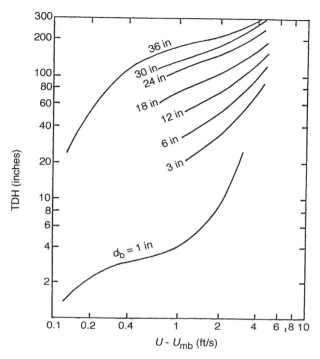

**Figure 5.12** Graph for determination of transport disengagement height after the method of Zenz (1983). Reproduced by permission. (Note: TDH and the bubble diameter at the bubble surface $d_b$ are given in inches (1 in = 25.4 mm)

$$\text{i.e. } R_i = -\frac{d}{dt}(M_B m_{Bi}) = K^*_{ih} A m_{Bi} \tag{5.38}$$

where $K^*_{ih}$ = elutriation rate constant (the entrainment flux at height $h$ above the bed surface for the solids of size $x_i$, when $m_{Bi} = 1.0$).

$M_B$ = total mass of solids in the bed

$A$ = area of bed surface

$m_{Bi}$ = fraction of the bed mass with size $x_i$ at time $t$.

For continuous operation, $m_{Bi}$ and $M_B$ are constant and so

$$R_i = K^*_{ih} A m_{Bi} \tag{5.39}$$

and total rate of entrainment, $R_T = \sum R_i = \sum K^*_{ih} A m_{Bi}$ \hfill (5.40)

The solids loading of size $x_i$ in the off-gases is $\rho_i = R_i/UA$ and the total solids loading leaving the freeboard is $\rho_T = \sum \rho_i$.

For batch operation, the rates of entrainment of each size range, the total entrainment rate and the particle size distribution of the bed change with time. The problem can best be solved by writing Equation (5.38) in finite increment form:

$$-\Delta(m_{Bi}M_B) = K_{ih}^* A m_{Bi}\Delta t \qquad (5.41)$$

where $\Delta(m_{Bi}M_B)$ is the mass of solids in size range $i$ entrained in time increment $\Delta t$.

$$\text{Then total mass entrained in time } \Delta t = \sum_{i=1}^{k}\{\Delta(m_{Bi}M_B)\} \qquad (5.42)$$

and mass of solids remaining in the bed at time

$$t + \Delta t = (M_B)_t - \sum_{i=1}^{k}\{\Delta(m_{Bi}M_B)_t\} \qquad (5.43)$$

(where subscript $t$ refers to the value at time $t$).

$$\text{Bed composition at time} t + \Delta t = (m_{Bi})_{t+\Delta t} = \frac{(m_{Bi}M_B)_t - [\Delta(m_{Bi}M_B)_t]}{(M_B)_t - \sum_{i=1}^{k}\{\Delta(m_{Bi}M_B)_t\}}$$
$$(5.44)$$

Solution of a batch entrainment problem proceeds by sequential application of Equations (5.41)–(5.44) for the required time period.

The elutriation rate constant $K_{ih}^*$ cannot be predicted from first principles and so it is necessary to rely on the available correlations which differ significantly in their predictions. Correlations are usually in terms of the carry-over rate above TDH, $K_{i\infty}^*$. Two of the more reliable correlations are given below:

Geldart *et al.* (1979): (for particles $> 100\ \mu m$ and $U > 1.2$ m/s)

$$\frac{K_{i\infty}^*}{\rho_g U} = 23.7 \exp\left[-5.4\frac{U_{Ti}}{U}\right] \qquad (5.45)$$

Zenz and Weil (1958) (for particles $< 100\ \mu m$ and $U < 1.2$ m/s)

$$\frac{K_{i\infty}^*}{\rho_g U} = \begin{cases} 1.26 \times 10^7 \left(\dfrac{U^2}{g x_i \rho_P^2}\right)^{1.88} & \text{when } \left(\dfrac{U^2}{g x_i \rho_P^2}\right) < 3 \times 10^{-4} \\[4mm] 4.31 \times 10^4 \left(\dfrac{U^2}{g x_i \rho_P^2}\right)^{1.18} & \text{when } \left(\dfrac{U^2}{g x_i \rho_P^2}\right) > 3 \times 10^{-4} \end{cases} \qquad (5.46)$$

## 5.7  HEAT TRANSFER IN FLUIDIZED BEDS

The transfer of heat between fluidized solids, gas and internal surfaces of equipment is very good. This makes for uniform temperatures and ease of control of bed temperature.

### 5.7.1  Gas-particle Heat Transfer

Gas to particle heat transfer coefficients are typically small, of the order of $5-20 \, \text{W/m}^2 \, \text{K}$. However, because of the very large heat transfer surface area provided by a mass of small particles ($1 \, \text{m}^3$ of $100 \, \mu\text{m}$ particles has a surface area of $60\,000 \, \text{m}^2$), the heat transfer between gas and particles is rarely limiting in fluid bed heat transfer. One of the most commonly used correlations for gas–particle heat transfer coefficient is that of Kunii and Levenspiel (1969):

$$Nu = 0.03 \, Re_{\text{p}}^{1.3} \qquad (Re_{\text{p}} < 50) \qquad (5.47)$$

where $Nu = $ Nusselt number $[h_{\text{gp}} x / k_{\text{g}}]$ and the single particle Reynolds number is based on the relative velocity between fluid and particle as usual.

Gas–particle heat transfer is relevant where a hot fluidized bed is fluidized by cold gas. The fact that particle–gas heat transfer presents little resistance in bubbling fluidized beds can be demonstrated by the following example:

Consider a fluidized bed of solids held at a constant temperature $T_{\text{s}}$. Hot fluidizing gas at temperature $T_{\text{g0}}$ enters the bed. At what distance above the distributor is the difference between the inlet gas temperature and the bed solids temperature reduced to half its original value?

Consider an element of the bed of height $\delta L$ at a distance $L$ above the distributor (Figure 5.13). Let the temperature of the gas entering this element be $T_{\text{g}}$ and the change in gas temperature across the element be $\delta T_{g}$. The particle temperature in the element is $T_{\text{s}}$.

The energy balance across the element gives

rate of heat loss by the gas = rate of heat transfer to the solids

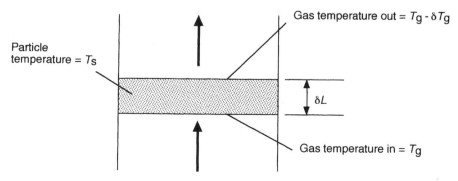

**Figure 5.13**  Analysis of gas–particle heat transfer in an element of a fluidized bed

$$\text{that is} - (C_g U \rho_g)\,dT_g = h_{gp} a (T_g - T_s)\,dL \tag{5.48}$$

where    $a$ = surface area of solids per unit volume of bed
   $C_g$ = specific heat capacity of the gas
   $\rho_p$ = particle density
   $h_{gp}$ = particle-to-gas heat transfer coefficient
   $U$ = superficial gas velocity.

Integrating with the boundary condition $T_g = T_{g0}$ at $L = 0$,

$$\ln\left[\frac{T_g - T_s}{T_{g0} - T_s}\right] = -\left[\frac{h_{gp} a}{U_{rel}\rho_g C_g}\right] L \tag{5.49}$$

The distance over which the temperature difference is reduced to half its initial value, $L_{0.5}$, is then

$$L_{0.5} = -\ln[0.5]\frac{C_g U_{rel}\rho_g}{h_{gp} a} = 0.693\frac{C_g U_{rel}\rho_g}{h_{gp} a} \tag{5.50}$$

For a bed of spherical particles of diameter $x$, the surface area per unit volume of bed, $a = 6(1 - \varepsilon)/x$ where $\varepsilon$ is the bed voidage.
   Using the correlation for $h_{gp}$ in Equation (5.47), then

$$L_{0.5} = 3.85\frac{\mu^{1.3} x^{0.7} C_g}{U_{rel}^{0.3}\rho_g^{0.3}(1 - \varepsilon)k_g} \tag{5.51}$$

As an example we will take a bed of particles of mean size 100 µm, particle density 2500 kg/m$^3$, fluidized by air of density 1.2 kg/m$^3$, viscosity $1.84 \times 10^{-5}$ Pa s, conductivity 0.0262 W/m K and specific heat capacity 1005 J/kg K.
   Using the Baeyens equation for $U_{mf}$ (Equation (5.11)), $U_{mf} = 9.3 \times 10^{-3}$ m/s. The relative velocity between particles and gas under fluidized conditions can be approximated as $U_{mf}/\varepsilon$ under these conditions.
   Hence, assuming a fluidized bed voidage of 0.47, $U_{rel} = 0.02$ m/s.
   Substituting these values in Equation (5.51), we find $L_{0.5} = 0.95$ mm. So, within 1 mm of entering the bed the difference in temperature between the gas and the bed will be reduced by half. Typically for particles less than 1 mm in diameter the temperature difference between hot bed and cold fluidizing gas would be reduced by half within the first 5 mm of the bed depth.

## 5.7.2   Bed-Surface Heat Transfer

In a bubbling fluidized bed the coefficient of heat transfer between bed and immersed surfaces (vertical bed walls or tubes) can be considered to be made up of three components which are approximately additive (Botterill, 1975).

$$\text{bed-surface heat transfer coefficient, } h = h_{pc} + h_{gc} + h_r$$

where $h_{pc}$ is the particle convective heat transfer coefficient and describes the heat transfer due to the motion of packets of solids carrying heat to and from the surface; $h_{gc}$ is the gas convective heat transfer coefficient describing the transfer of heat by motion of the gas between the particles; $h_r$ is the radiant heat transfer coefficient. Figure 5.14, after Botterill (1986), gives an indication of the range of bed-surface heat transfer coefficients and the effect of particle size on the dominant heat transfer mechanism.

Particle convective heat transfer: On a volumetric basis the solids in the fluidized bed have about one thousand times the heat capacity of the gas and so, since the solids are continuously circulating within the bed, they transport the heat around the bed. For heat transfer between the bed and a surface the limiting factor is the gas conductivity, since all the heat must be transferred through a gas film between the particles and the surface (Figure 5.15). The particle–surface contact area is too small to allow significant heat transfer. Factors affecting the gas film thickness or the gas conductivity will therefore influence the heat transfer under particle convective conditions. Decreasing particle size, for example, decreases the mean gas film thickness and so improves $h_{pc}$. However, reducing particle size into the Group C range will reduce particle mobility and so reduce particle convective heat transfer. Increasing gas temperature increases gas conductivity and so improves $h_{pc}$.

Particle convective heat transfer is dominant in Group A and B powders. Increasing gas velocity beyond minimum fluidization improves particle circulation and so increases particle convective heat transfer. The heat transfer coefficient increases with fluidizing velocity up to a broad maximum $h_{max}$ and then declines as the heat transfer surface becomes blanketed by bubbles. This is shown in Figure 5.16 for powders in Groups A, B and D. The maximum in

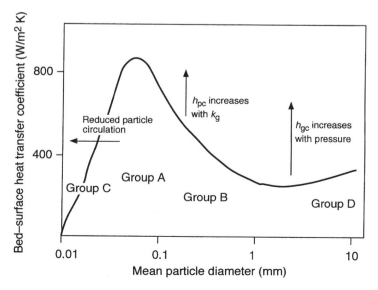

**Figure 5.14** Range of bed–surface heat transfer coefficients

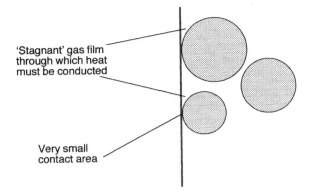

**Figure 5.15** Heat transfer from bed particles to an immersed surface

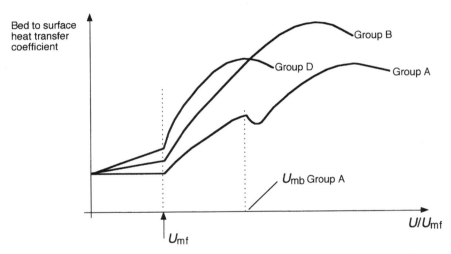

**Figure 5.16** Effect of fluidizing gas velocity on bed–surface heat transfer coefficient in a fluidized bed

$h_{pc}$ occurs relatively closer to $U_{mf}$ for Group B and D powders since these powders give rise to bubbles at $U_{mf}$ and the size of these bubbles increases with increasing gas velocity. Group A powders exhibit a non-bubbling fluidization between $U_{mf}$ and $U_{mb}$ and achieve a maximum stable bubble size.

Botterill (1986) recommends the Zabrodsky (1966) correlation for $h_{max}$ for Group B powders:

$$h_{max} = 35.8 \frac{k_g^{0.6} \rho_p^{0.2}}{x^{0.36}} \qquad W/m^2K \qquad (5.52)$$

and the correlation of Khan et al. (1978) for Group A powders:

$$Nu_{max} = 0.157 Ar^{0.475} \qquad (5.53)$$

Gas convective heat transfer is not important in Group A and B powders where the flow of interstitial gas is laminar but becomes significant in Group D powders, which fluidize at higher velocities and give rise to transitional or turbulent flow of interstitial gas. Botterill suggests that the gas convective mechanism takes over from particle convective heat transfer as the dominant mechanism at $Re_{mf} \approx 12.5$ ($Re_{mf}$ is the Reynolds number at minimum fluidization and is equivalent to an Archimedes number $Ar \approx 26\,000$). In gas convective heat transfer the gas specific heat capacity is important as the gas transports the heat around. Gas specific heat capacity increases with increasing pressure and in conditions where gas convective heat transfer is dominant, increasing operating pressure gives rise to an improved heat transfer coefficient $h_{gc}$. Botterill (1986) recommends the correlations of Baskakov and Suprun (1972) for $h_{gc}$:

$$Nu_{gc} = 0.0175 Ar^{0.46} Pr^{0.33} \qquad \text{(for } U > U_m\text{)} \qquad (5.54)$$

$$Nu_{gc} = 0.0175 Ar^{0.46} Pr^{0.33} \left(\frac{U}{U_m}\right)^{0.3} \qquad \text{(for } U_{mf} < U < U_m\text{)} \qquad (5.55)$$

where $U_m$ is the superficial velocity corresponding to the maximum overall bed heat transfer coefficient.

For temperatures beyond 600°C radiative heat transfer plays an increasing role and must be accounted for in calculations. The reader is referred to Botterill (1986) or Kunii and Levenspiel (1990) for treatment of radiative heat transfer or for a more detailed look at heat transfer in fluidized beds.

## 5.8 APPLICATIONS OF FLUIDIZED BEDS

### 5.8.1 Physical Processes

Physical processes which use fluidized beds include drying, mixing, granulation, coating, heating and cooling. All these processes take advantage of the excellent mixing capabilities of the fluid bed. Good solids mixing gives rise to good heat transfer, temperature uniformity and ease of process control. One of the most important applications of the fluidized bed is to the drying of solids. Fluidized beds are currently used commercially for drying such materials as crushed minerals, sand, polymers, pharmaceuticals, fertilizers and crystalline products. The reasons for the popularity of fluidized bed drying are:

- The dryers are compact, simple in construction and of relatively low capital cost.

- The absence of moving parts, other than the feeding and discharge devices, leads to reliable operation and low maintenance.

- The thermal efficiency of these dryers is relatively high.

- Fluidized bed dryers are gentle in the handling of powders and this is useful when dealing with friable materials.

Fluidized bed granulation is dealt with in Chapter 11 and mixing is covered in Chapter 9. Fluidized beds are often used to cool particulate solids following a reaction. Cooling may be by fluidizing air alone or by the use of cooling water passing through tubes immersed in the bed (see Figure 5.17 for example). Fluidized beds are used for coating particles in the pharmaceutical and agricultural industries. Metal components may be plastic coated by dipping them hot into an air-fluidized bed of powdered thermosetting plastic.

### 5.8.2   Chemical Processes

The gas fluidized bed is a good medium in which to carry out a chemical reaction involving a gas and a solid. Advantages of the fluidized bed for chemical reaction include:

- The gas–solids contacting is generally good.

- The excellent solids circulation within the bed promotes good heat transfer between bed particles and the fluidizing gas and between the bed and heat transfer surfaces immersed in the bed.

- This gives rise to near isothermal conditions even when reactions are strongly exothermic or endothermic.

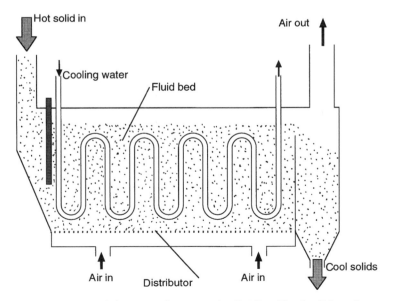

**Figure 5.17**   Schematic diagram of a fluidized bed solid cooler

- The good heat transfer also gives rise to ease of control of the reaction.

- The fluidity of the bed makes for ease of removal of solids from the reactor.

However, it is far from ideal; the main problems arise from the two phase (bubbles and fluidized solids) nature of such systems. This problem is particularly acute where the bed solids are the catalyst for a gas-phase reaction. In such a case the ideal fluidized bed chemical reactor would have excellent gas–solid contacting, no gas by-passing and no back-mixing of the gas against the main direction of flow. In a bubbling fluidized bed the gas by-passes the solids by passing through the bed as bubbles. This means that unreacted reactants appear in the product. Also, gas circulation patterns within a bubbling fluidized bed are such that products are back-mixed and may undergo undesirable secondary reactions. These problems lead to serious practical difficulties particularly in the scaling-up of a new fluidized bed process from pilot plant to full industrial scale. This subject is dealt with in more detail in references: Kunii and Levenspiel (1990), Geldart (1986), Davidson and Harrison (1971).

Figure 5.18 is a schematic diagram of one type of fluid catalytic cracking (FCC) unit, a celebrated example of fluidized bed technology for breaking down large molecules in crude oil to small molecules suitable for gasoline etc. Other examples of the application of fluidized bed technology to different kinds of chemical reaction are shown in Table 5.2.

## 5.9 A SIMPLE MODEL FOR THE BUBBLING FLUIDIZED BED REACTOR

In general, models for the fluidized bed reactor consider:

- the division of gas between the bubble phase and particulate phase;
- the degree of mixing in the particulate phase;
- the transfer of gas between the phases.

It is outside the scope of this chapter to review in detail the models available for the fluidized bed as a reactor. However, in order to demonstrate the key components of such models, we will use the simple model of Orcutt et al. (1962). Although simple, this model allows the key features of a fluidized bed reactor for gas-phase catalytic reaction to be explored.

The approach assumes the following:

- Original two-phase theory applies.
- Perfect mixing takes place in the particulate phase.
- There is no reaction in the bubble phase.

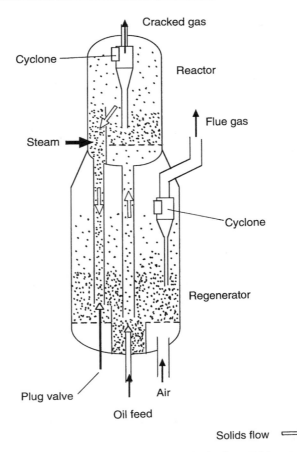

**Figure 5.18**   Kellogg's Model A Orthoflow FCC unit

**Table 5.2**   Summary of the types of gas–solid chemical reactions employing fluid-ization

| Type | Example | Reasons for using a fluidized bed |
|------|---------|-----------------------------------|
| Homogeneous gas-phase reactions | Ethylene hydrogenation | Rapid heating of entering gas. Uniform controllable temperature |
| Heterogeneous non-catalytic reactions | Sulphide ore roasting, combustion | Ease of solids handling. Temperature uniformity. Good heat transfer |
| Heterogeneous catalytic reactions | Hydrocarbon cracking, phthalic anhydride, acrylonitrile | Ease of solids handling. Temperature uniformity. Good heat transfer |

The model is one-dimensional and assumes steady state. The structure of the model is shown diagramatically in Figure 5.19. The following is the nomenclature used:

$C_0$ = concentration of reactant at distributor;

$C_p$ = concentration of reactant in the particulate phase;

$C_B$ = concentration of the reactant in the bubble phase at height $h$ above the distributor;

$C_{BH}$ = concentration of reactant leaving the bubble phase;

$C_H$ = concentration of reactant leaving the reactor.

In steady state, the concentration of reactant in the particulate phase is constant throughout the particulate phase because of the assumption of perfect mixing in the particulate phase. Throughout the bed gaseous reactant is assumed to pass between particulate phase and bubble phase.

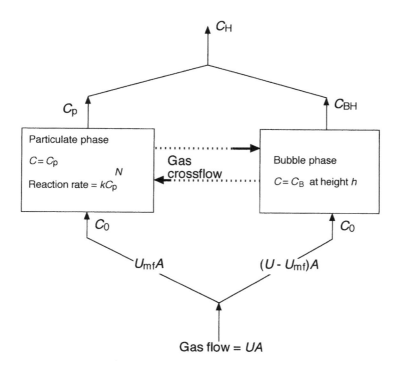

**Figure 5.19**   Schematic diagram of the Orcutt fluidized bed reactor model

The overall mass balance on the reactant is

$$
\begin{bmatrix} \text{molar flow of} \\ \text{reactant in} \end{bmatrix} = \begin{bmatrix} \text{molar flow out} \\ \text{in the bubble phase} \end{bmatrix} + \begin{bmatrix} \text{molar flow out in} \\ \text{the particulate phase} \end{bmatrix} + \begin{bmatrix} \text{rate of} \\ \text{conversion} \end{bmatrix}
$$
$$\quad\quad (1) \qquad\qquad\qquad (2) \qquad\qquad\qquad\qquad (3) \qquad\qquad\qquad (4)$$

$$(5.56)$$

Term (1) $= UAC_0$

Term (2): molar reactant flow in bubble phase changes with height $L$ above the distributor as gas is exchanged with the particulate phase. Consider an element of bed of thickness $\delta L$ at a height $L$ above the distributor. In this element:

$$
\begin{bmatrix} \text{rate of increase of} \\ \text{reactant in bubble phase} \end{bmatrix} = \begin{bmatrix} \text{rate of transfer of} \\ \text{reactant from particulate phase} \end{bmatrix}
$$

$$\text{i.e. } (U - U_{mf})A\delta C_B = -K_C(\varepsilon_B A\delta L)(C_B - C_p) \qquad (5.57)$$

in the limit as $\delta L \rightarrow 0$,

$$\frac{dC_B}{dL} = -\frac{K_C\varepsilon_B(C_B - C_p)}{(U - U_{mf})} \qquad (5.58)$$

Where $K_C$ is the mass transfer coefficient per unit bubble volume and $\varepsilon_B$ is the bubble fraction. Integrating with the boundary condition that $C_B = C_0$ at $L = 0$:

$$C_B = C_p + (C_0 - C_p)\exp\left[-\frac{K_C L}{U_B}\right] \qquad (5.59)$$

since $\varepsilon_B = (U - U_{mf})/U_B$ (Equation (5.28))

At the surface of the bed, $L = H$ and so the reactant concentration in the bubble phase at the bed surface is given by

$$C_{BH} = C_p + (C_0 - C_p)\exp\left[-\frac{K_C H}{U_B}\right] \qquad (5.60)$$

and so,  Term (2) $= C_{BH}(U - U_{mf})A$

Term (3) $= U_{mf}AC_p$

Term (4) $=$

For a reaction which is $j$th order in the reactant under consideration,

$$\begin{bmatrix} \text{molar rate of conversion} \\ \text{per unit volume of solids} \end{bmatrix} = kC_P^j$$

where $k$ = reaction rate per unit volume of solids

Therefore,

$$\begin{bmatrix} \text{molar rate of} \\ \text{conversion in bed} \end{bmatrix} = \begin{bmatrix} \text{molar rate of} \\ \text{conversion per unit} \\ \text{volume of solids} \end{bmatrix} \times \begin{bmatrix} \text{volume of solids} \\ \text{per unit volume of} \\ \text{particulate phase} \end{bmatrix}$$

$$\times \begin{bmatrix} \text{volume of particulate} \\ \text{phase per unit} \\ \text{volume of bed} \end{bmatrix} \times \begin{bmatrix} \text{volume} \\ \text{of bed} \end{bmatrix}$$

hence, Term (4),

$$\begin{bmatrix} \text{molar rate of} \\ \text{conversion in bed} \end{bmatrix} = kC_p^j(1 - \varepsilon_p)(1 - \varepsilon_B)AH \tag{5.61}$$

$\varepsilon_p$ = particulate phase voidage
substituting these expressions for the terms (1)–(4), the mass balance becomes

$$UAC_0 = \left\{ C_p + (C_0 - C_p)\exp\left[-\frac{K_C H}{U_B}\right] \right\}(U - U_{mf})A + U_{mf}AC_p$$

$$+ kC_p^j(1 - \varepsilon_p)(1 - \varepsilon_B)AH \tag{5.62}$$

From this mass balance $C_p$ may be found. The reactant concentration leaving the reactor $C_H$ is then calculated from the reactant concentrations and gas flows through the bubble and particulate phases:

$$C_H = \frac{U_{mf}C_p + (U - U_{mf})C_{BH}}{U} \tag{5.63}$$

In the case of a first order reaction ($j = 1$), solving the mass balance for $C_p$ gives

$$C_p = \frac{C_0[U - (U - U_{mf})e^{-\chi}]}{kH_{mf}(1 - \varepsilon_p) + [U - (U - U_{mf})e^{-\chi}]} \tag{5.64}$$

where $\chi = K_C H/U_B$, equivalent to a number of mass transfer units for gas exchange between the phases. $\chi$ is related to bubble size and correlations are available. Generally $\chi$ decreases as bubble size increases and so small bubbles are preferred.

Thus from Equation (5.63) and (5.64), we obtain an expression for the conversion in the reactor:

$$1 - \frac{C_H}{C_0} = (1 - \beta e^{-\chi}) - \frac{(1 - \beta e^{-\chi})^2}{\dfrac{kH_{mf}(1 - \varepsilon_p)}{U} + (1 - \beta e^{-\chi})} \tag{5.65}$$

where $\beta = (U - U_{mf})/U$, the fraction of gas passing through the bed as bubbles. It is interesting to note that although the two-phase theory does not always hold, Equation (5.65) often holds with $\beta$ still the fraction of gas passing through the bed as bubbles, but not equal to $(U - U_{mf})/U$.

Readers interested in reactions of order different from unity, solids reactions and more complex reactor models for the fluidized bed, are referred to Kunii and Levenspiel (1990).

Although the Orcutt model is simple, it does allow us to explore the effects of operating conditions, reaction rate and degree of interphase mass transfer on performance of a fluidized bed as a gas-phase catalytic reactor. Figure 5.20 shows the variation of conversion with reaction rate (expressed as $kH_{mf}(1 - \varepsilon_p)/U$) with excess gas velocity (expressed as $\beta$) calculated using Equation (5.65) for a first-order reaction.

Noting that the value of $\chi$ is dictated mainly by the bed hydrodynamics, we see that

- *For slow reactions*, overall conversion is insensitive to bed hydrodynamics and so reaction rate k is the rate controlling factor.

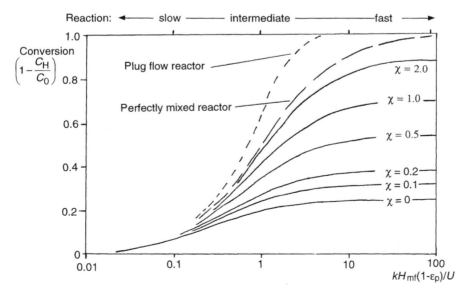

**Figure 5.20**  Conversion as a function of reaction rate and interphase mass transfer for $\beta = 0.75$ for a first order gas phase catalytic reaction (based on Equation (5.65))

- *For intermediate reactions,* both reaction rate and bed hydrodynamics affect the conversion.

- *For fast reactions,* the conversion is determined by the bed hydrodynamics.

These results are typical for a gas-phase catalytic reaction in a fluidized bed.

## 5.10   SOME PRACTICAL CONSIDERATIONS

### 5.10.1   Gas Distributor

The distributor is a device designed to ensure that the fluidizing gas is always evenly distributed across the cross-section of the bed. It is a critical part of the design of a fluidized bed system. Good design is based on achieving a pressure drop which is a sufficient fraction of the bed pressure drop. Readers are referred to Geldart (1986) for guidelines on distributor design. Many operating problems can be traced back to poor distributor design. Some distributor designs in common use are shown in Figure 5.21.

### 5.10.2   Loss of Fluidizing Gas

Loss of fluidizing gas will lead to collapse of the fluidized bed. If the process involves the emission of heat then this heat will not be dissipated as well from

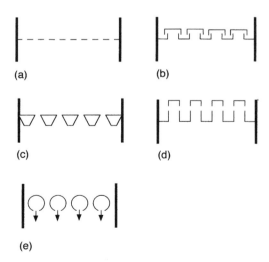

**Figure 5.21**   Some distributor designs in common use (a) drilled plate; (b) cap design; (c) continuous horizontal slots; (d) stand pipe design; (e) sparge tubes with holes pointing downwards

the packed bed as it was from the fluidized bed. The consequences of this should be considered at the design stage.

### 5.10.3   Erosion

All parts of the fluidized bed unit are subject to erosion by the solid particles. Heat transfer tubes within the bed or freeboard are particularly at risk and erosion here may lead to tube failure. Erosion of the distributor may lead to poor fluidization and areas of the bed becoming deaerated.

### 5.10.4   Loss of Fines

Loss of fine solids from the bed reduces the quality of fluidization and reduces the area of contact between the solids and the gas in the process. In a catalytic process this means lower conversion.

### 5.10.5   Cyclones

Cyclone separators are often used in fluidized beds for separating entrained solids from the gas stream (see Chapter 7). Cyclones installed within the fluidized bed vessel would be fitted with a dip-leg and seal in order to prevent gas entering the solids exit. Fluidized systems may have two or more stages of cyclone in series in order to improve separation efficiency. Cyclones are also the subject of erosion and must be designed to cope with this.

### 5.10.6   Solids Feeders

Various devices are available for feeding solids into the fluidized bed. The choice of device depends largely on the nature of the solids feed. Screw conveyors, spray feeders and pneumatic conveying are in common use.

## 5.11 WORKED EXAMPLES

### WORKED EXAMPLE 5.1

3.6 kg of solid particles of density 2590 kg/m³ and surface-volume mean size 748 μm form a packed bed of height 0.475 m in a circular vessel of diameter 0.0757 m. Water of density 1000 kg/m³ and viscosity 0.001 Pa s is passed upwards through the bed. Calculate (a) the bed pressure drop at incipient fluidization, (b) the superficial liquid velocity at incipient fluidization, (c) the mean bed voidage at a superficial liquid velocity of 1.0 cm/s, (d) the bed height at this velocity and (e) the pressure drop across the bed at this velocity.

## Solution

(a) Applying Equation (5.24) to the packed bed, we find the packed bed voidage:

$$\text{mass of solids} = 3.6 = (1 - \varepsilon) \times 2590 \times \frac{\pi(0.0757)^2}{4} \times 0.475$$

hence, $\varepsilon = 0.3498$

Frictional pressure drop across the bed when fluidized:

$$(-\Delta p) = \frac{\text{weight of particles} - \text{upthrust on particles}}{\text{cross-sectional area}}$$

$$(-\Delta p) = \frac{Mg - Mg(\rho_f/\rho_p)}{A} \quad \text{(since upthrust} = \text{weight of fluid displaced by particles)}$$

$$\text{Hence, } (-\Delta p) = \frac{Mg}{A}\left(1 - \frac{\rho_f}{\rho_p}\right) = \frac{3.6 \times 9.81}{4.50 \times 10^{-3}}\left(1 - \frac{1000}{2590}\right) = 4817 \text{ Pa}$$

(b) Assuming that the voidage at the onset of fluidization is equal to the voidage of the packed bed, we use the Ergun equation to express the relationship between packed pressure drop and superficial liquid velocity:

$$\frac{(-\Delta p)}{H} = 3.55 \times 10^7 U^2 + 2.648 \times 10^6 U$$

Equating this expression for pressure drop across the packed bed to the fluidized bed pressure drop, we determine superficial fluid velocity at incipient fluidization, $U_{mf}$.

$$U_{mf} = 0.365 \text{ cm/s}$$

(c) The Richardson–Zaki equation (Equation 5.21) allows us to estimate the expansion of a liquid fluidized bed.

$$U = U_T \varepsilon^n \hspace{3cm} \text{(Equation (5.21))}$$

Using the method given in Chapter 1, we determine the single particle terminal velocity, $U_T$.

$$Ar = 6527.9; \quad C_D Re_p^2 = 8704; \quad Re_p = 90; \quad U_T = 0.120 \text{ m/s}$$

At this value of Reynolds number, the flow is intermediate between viscous and inertial, and so we must use the correlation of Khan and Richardson (Equation (2.25)) to determine the exponent $n$ in Equation (5.21):

$$\text{With } Ar = 6527.9, \ n = 3.202$$

Hence from Equation (5.21), $\varepsilon = 0.460$ when $U = 0.01$ m/s
Mean bed voidage is 0.460 when the superficial liquid velocity is 1 cm/s.

(d) From Equation (5.25), we now determine the mean bed height at this velocity:

$$\text{Bed height (at } U = 0.01 \text{ m/s)} = \frac{(1 - 0.3498)}{(1 - 0.460)} 0.475 = 0.572 \text{ m}$$

(e) The frictional pressure drop across the bed remains essentially constant once the bed is fluidized. Hence at a superficial liquid velocity of 1 cm/s the frictional pressure drop across the bed is 4817 Pa.

However, the measured pressure drop across the bed will include the hydro-static head of the liquid in the bed. Applying the mechanical energy equation between the bottom (1) and the top (2) of the fluidized bed:

$$\frac{p_1 - p_2}{\rho_f g} + \frac{U_1^2 - U_2^2}{2g} + (z_1 - z_2) = \text{friction head loss} = \frac{4817}{\rho_f g}$$

$$U_1 = U_2; \qquad z_1 - z_2 = -H = -0.572 \text{ m.}$$

Hence, $p_1 - p_2 = 10\,428$ Pa.

## WORKED EXAMPLE 5.2

A powder having the size distribution given below and a particle density of 2500 kg/m$^3$ is fed into a fluidized bed of cross-sectional area 4 m$^2$ at a rate of 1.0 kg/s.

| Size range number ($i$) | Size range (μm) | Mass fraction in feed |
|---|---|---|
| 1 | 10–30 | 0.20 |
| 2 | 30–50 | 0.65 |
| 3 | 50–70 | 0.15 |

The bed is fluidized using air of density 1.2 kg/m$^3$ at a superficial velocity of 0.25 m/s. Processed solids are continuously withdrawn from the base of the fluidized bed in order to maintain a constant bed mass. Solids carried over with the gas leaving the vessel are collected by a bag filter operating at 100% total efficiency. None of the solids caught by the filter are returned to the bed. Assuming that the fluidized bed is well-mixed and that the freeboard height is greater than the transport disengagement height under these conditions, calculate at equilibrium:

(a)  the flowrate of solids entering the filter bag;

(b)  the size distribution of the solids in the bed;

(c)  the size distribution of the solids entering the filter bag;

(d)  the rate of withdrawal of processed solids from the base of the bed;

(e)  the solids loading in the gas entering the filter.

## Solution

First calculate the elutriation rate constants for the three size ranges under these conditions from the Zenz and Weil correlation (Equation (5.46)). The value of particle size $x$ used in the correlation is the arithmetic mean of each size range:

$$x_1 = 20 \times 10^{-6} \text{ m}; \ x_2 = 40 \times 10^{-6} \text{ m}; \ x_3 = 60 \times 10^{-6} \text{ m}$$

With $U = 0.25$ m/s, $\rho_p = 2500$ kg/m$^3$ and $\rho_f = 1.2$ kg/m$^3$

$$K_{1\infty}^* = 3.21 \times 10^{-2} \text{ kg/m}^2 \text{ s};$$

$$K_{2\infty}^* = 8.74 \times 10^{-3} \text{ kg/m}^2 \text{ s};$$

$$K_{3\infty}^* = 4.08 \times 10^{-3} \text{ kg/m}^2 \text{ s}$$

Referring to Figure 5.W2.1 the overall and component material balances over the fluidized bed system are:

$$\text{Overall balance:} \qquad F = Q + R \qquad\qquad (5\text{W}2.1)$$

$$\text{Component balance:} \qquad Fm_{F_i} = Qm_{Q_i} + Rm_{R_i} \qquad\qquad (5\text{W}2.2)$$

where $F$, $Q$ and $R$ are the mass flowrates of solids in the feed, withdrawal and filter discharge respectively and $m_{F_i}$, $m_{Q_i}$ and $m_{R_i}$ are the mass fractions of solids in size range $i$ in the feed, withdrawal and filter discharge respectively.

From Equation (5.39) the entrainment rate of size range $i$ at the gas exit from the freeboard is given by

$$R_i = Rm_{R_i} = K_{i\infty}^* Am_{B_i} \qquad\qquad (5\text{W}2.3)$$

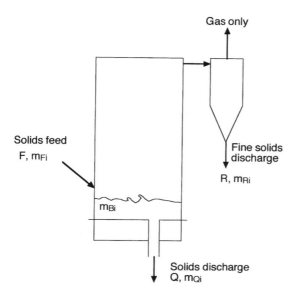

**Figure 5.W2.1** Schematic diagram showing solids flows and size distributions for the fluidized bed in worked example 5.2

and

$$R = \sum R_i = \sum R m_{R_i} \qquad (5W2.4)$$

Combining these equations with the assumption that the bed is well mixed $(m_{Q_i} = m_{B_i})$,

$$m_{B_i} = \frac{F m_{F_i}}{F - R + K^*_{i\infty} A} \qquad (5W2.5)$$

Now both $m_{B_i}$ and $R$ are unknown. However, noting that $\sum m_{B_i} = 1$, we have

$$\frac{1.0 \times 0.2}{1.0 - R + (3.21 \times 10^{-2} \times 4)} + \frac{1.0 \times 0.65}{1.0 - R + (8.74 \times 10^{-3} \times 4)} + \frac{1.0 \times 0.15}{1.0 - R + (4.08 \times 10^{-3} \times 4)} = 1.0$$

Solving for $R$ by trial and error, $R = 0.05$ kg/s

(b) Substituting $R = 0.05$ kg/s in Equation (5W2.5), $m_{B_1} = 0.1855$; $m_{B_2} = 0.6599$ and $m_{B_3} = 0.1552$
Therefore size distribution of bed:

| Size range number ($i$) | Size range (µm) | Mass fraction in bed |
|---|---|---|
| 1 | 10–30 | 0.1855 |
| 2 | 30–50 | 0.6599 |
| 3 | 50–70 | 0.1552 |

(c) From Equation (5W2.3), knowing $R$ and $m_{B_i}$, we can calculate $m_{R_i}$:

$$m_{R_i} = \frac{K^*_{1\infty} A m_{B_1}}{R} = \frac{3.21 \times 10^{-2} \times 4 \times 0.1855}{0.05} = 0.476$$

similarly, $m_{R_2} = 0.4614$; $m_{R_3} = 0.0506$
Therefore size distribution of solids entering filter:

| Size range number ($i$) | Size range (µm) | Mass fraction entering filter |
|---|---|---|
| 1 | 10–30 | 0.476 |
| 2 | 30–50 | 0.4614 |
| 3 | 50–70 | 0.0506 |

(d) From Equation (5W2.1), the rate of withdrawal of solids from the bed, $Q = 0.95$ kg/s

(e) Solids loading for gas entering the filter,

$$\frac{\text{mass flow of solids}}{\text{volume flow of gas}} = \frac{R}{UA} = 0.05 \text{ kg/m}^3$$

## WORKED EXAMPLE 5.3

A gas phase catalytic reaction is performed in a fluidized bed operating at a superficial gas velocity of 0.3 m/s. For this reaction under these conditions it is known that the reaction is first order in reactant A. Given the following information,

- bed height at incipient fluidization = 1.5 m;

- operating mean bed height = 1.65 m;

- voidage at incipient fluidization = 0.47;

- reaction rate constant = 75.47 (per unit volume of solids);

- mean bubble rise velocity = 0.111 m/s;

- mass transfer coefficient between bubbles and emulsion = 0.1009 (based on unit bubble volume).

use the reactor model of Orcutt *et al.* to determine

(a)  the conversion of reactant A;

(b)  the effect on the conversion found in (a) of reducing the inventory of catalyst by one half;

(c)  the effect on the conversion found in part (a) of halving the bubble size (assuming the interphase mass transfer coefficient is inversely proportional to the square root of the bubble diameter).

Discuss your answers to parts (b) and (c) and state which mechanism is controlling conversion in the reactor.

## Solution

(a) From Section 5.9 the model of Orcutt *et al.* gives for a first order reaction:

$$\text{conversion, } 1 - \frac{C_H}{C_0} = (1 - \beta e^{-\chi}) - \frac{(1 - \beta e^{-\chi})^2}{\dfrac{kH_{mf}(1 - \varepsilon_p)}{U} + (1 - \beta e^{-\chi})} \qquad \text{(Equation (5.65))}$$

where

$$\chi = \frac{K_C H}{U_B} \text{ and } \beta = (U - U_{mf})/U$$

From the information given in the question,

$$K_C = 0.1009, \ U_B = 0.1 \text{ m/s}, \ U = 0.3 \text{ m/s}, \ U_{mf} = 0.033 \text{ m/s},$$

$$H = 1.65 \text{ m}, \ H_{mf} = 1.5 \text{ m}, \ k = 75.47.$$

Hence, $\chi = 1.5, \beta = 0.89$ and $kH_{mf}(1 - \varepsilon_p)/U = 200$ (assuming $\varepsilon_p = \varepsilon_{mf}$)
So, from Equation (5.65), conversion $= 0.798$.

(b) If the inventory of solids in the bed is halved, both the operating bed height $H$ and the height at incipient fluidization $H_{mf}$ are halved. Thus, assuming all else remains constant, under the new conditions

$$\chi = 0.75, \ \beta = 0.89 \text{ and } kH_{mf}(1 - \varepsilon_p)/U = 100$$

and so the new conversion $= 0.576$

(c) If the bubble size is halved and $K_C$ is proportional to $1/\sqrt{(\text{bubble diameter})}$,

$$\text{new } K_C = 1.414 \times 0.1009 = 0.1427$$

Hence, $\chi = 2.121$, giving conversion $= 0.889$.

(d) Comparing the conversion achieved in part (c) with that achieved in part (a), we see that improving interphase mass transfer has a significant effect on the conversion. We may also note that doubling the reaction rate (say by increasing the reactor temperature) and keeping everything else constant has a negligible effect on the conversion achieved in part (a). We conclude, therefore, that under these conditions the transfer of gas between bubble phase and emulsion phase controls the conversion.

## EXERCISES

**5.1** A packed bed of solid particles of density 2500 kg/m³, occupies a depth of 1 m in a vessel of cross-sectional area 0.04 m². The mass of solids in the bed is 50 kg and the surface-volume mean diameter of the particles is 1 mm. A liquid of density 800 kg/m³ and viscosity 0.002 Pa s flows upwards through the bed.

(a)  Calculate the voidage (volume fraction occupied by voids) of the bed.

(b)  Calculate the pressure drop across the bed when the volume flow rate of liquid is 1.44 m³/h.

(c)  Calculate the pressure drop across the bed when it becomes fluidized. (Answer: (a) 0.5; (b) 6560 Pa; (c) 8338 Pa)

**5.2** 130 kg of uniform spherical particles with a diameter of 50 μm and particle density 1500 kg/m³ are fluidized by water (density 1000 kg/m³, viscosity 0.001 Pa s.) in a circular bed of cross-sectional area 0.2 m². The single particle terminal velocity of the particles is 0.68 mm/s and the voidage at incipient fluidization is known to be 0.47.

(a) Calculate the bed height at incipient fluidization.

(b) Calculate the mean bed voidage when the liquid flow rate is $2 \times 10^{-5}$ m$^3$/s.

(Answer: (a) 0.818 m; (b) 0.6622)

**5.3** 130 kg of uniform spherical particles with a diameter of 60 μm and particle density 1500 kg/m$^3$ are fluidized by water (density 1000 kg/m$^3$, viscosity 0.001 Pa s) in a circular bed of cross-sectional area 0.2 m$^2$. The single particle terminal velocity of the particles is 0.98 mm/s and the voidage at incipient fluidization is known to be 0.47.

(a) Calculate the bed height at incipient fluidization.

(b) Calculate the mean fluidized bed voidage when the liquid flow rate is $2 \times 10^{-5}$ m$^3$/s.

(Answer: (a) 0.818 m; (b) 0.6121)

**5.4** A packed bed of solid particles of density 2500 kg/m$^3$, occupies a depth of 1 m in a vessel of cross-sectional area 0.04 m$^2$. The mass of solids in the bed is 45 kg and the surface-volume mean diameter of the particles is 1 mm. A liquid of density 800 kg/m$^3$ and viscosity 0.002 Pa s flows upwards through the bed.

(a) Calculate the voidage (volume fraction occupied by voids) of the bed.

(b) Calculate the pressure drop across the bed when the volume flow rate of liquid is 1.44 m$^3$/h.

(c) Calculate the pressure drop across the bed when it becomes fluidized.

(Answer: (a) 0.55; (b) 4030 Pa; (c) 7505 Pa.)

**5.5** 12 kg of spherical resin particles of density 1200 kg/m$^3$ and uniform diameter 70 μm are fluidized by water (density 1000 kg/m$^3$ and viscosity 0.001 Pa s) in a vessel of diameter 0.3 m and form an expanded bed of height 0.25 m.

(a) Calculate the difference in pressure between the base and the top of the bed.

(b) If the flow rate of water is increased to 7 cm$^3$/s, what will be the resultant bed height and bed voidage (liquid volume fraction)?

State and justify the major assumptions.

(Answer: (a) Frictional pressure drop = 277.5 Pa; Pressure difference = 2730 Pa; (b) height = 0.465 m; voidage = 0.696)

**5.6** A packed bed of solids of density 2000 kg/m$^3$ occupies a depth of 0.6 m in a cylindrical vessel of inside diameter 0.1 m. The mass of solids in the bed is 5 kg and the surface-volume mean diameter of the particles is 300 µm. Water (density 1000 kg/m³ and viscosity 0.001 Pa s) flows upwards through the bed.

(a) What is the voidage of the packed bed?

(b) Use a force balance over the bed to determine the bed pressure drop when fluidized.

(c) Hence, assuming laminar flow and that the voidage at incipient fluidization is the same as the packed bed voidage, determine the minimum fluidization velocity. Verify the assumption of laminar flow.

(Answer: (a) 0.4692; (b) 3124 Pa; (c) 1.145 mm/s)


**5.7** A packed bed of solids of density 2000 kg/m³ occupies a depth of 0.5 m in a cylindrical vessel of inside diameter 0.1 m. The mass of solids in the bed is 4 kg and the surface-volume mean diameter of the particles is 400 µm. Water (density 1000 kg/m³ and viscosity 0.001 Pa s) flows upwards through the bed.

(a) What is the voidage of the packed bed?

(b) Use a force balance over the bed to determine the bed pressure drop when fluidized.

(c) Hence, assuming laminar flow and that the voidage at incipient fluidization is the same as the packed bed voidage, determine the minimum fluidization velocity. Verify the assumption of laminar flow.

(Answer: (a) 0.4907; (b) 2498 Pa; (c) 2.43 mm/s)


**5.8** By applying a force balance, calculate the incipient fluidizing velocity for a system with particles of particle density 5000 kg/m³ and mean volume diameter 100 µm and a fluid of density 1.2 kg/m³ and viscosity $1.8 \times 10^{-5}$ Pa s. Assume that the voidage at incipient fluidization is 0.5.
   If in the above example the particle size is changed to 2 mm, what is $U_{mf}$?

(Answer: 0.045 m/s; 2.26 m/s)


**5.9** A powder of mean sieve size 60 µm and particle density 1800 kg/m³ is fluidized by air of density 1.2 kg/m³ and viscosity $1.84 \times 10^{-5}$ Pa s in a circular vessel of diameter 0.5 m. The mass of powder charged to the bed is 240 kg and the volume flowrate of air to the bed is 140 m³/h. It is known that the average bed voidage at incipient fluidization is 0.45 and correlation reveals that the average bubble rise velocity under the conditions in question is 0.8 m/s. Estimate:

(a) the minimum fluidization velocity, $U_{mf}$

(b) the bed height at incipient fluidization

(c) the visible bubble flowrate

(d)  the bubble fraction

(e)  the particulate phase voidage

(f)  the mean bed height

(g)  the mean bed voidage

(Answer: (a) Ergun 0.38 cm/s; Wen and Yu 0.213 cm/s; (b) 1.24 m; (c) 0.038 m³/s; (d) 0.245; (e) 0.45; (f) 1.64 m; (g) 0.585)

**5.10** A batch fluidized bed process has an initial charge of 2000 kg of solids of particle density 1800 kg/s³ and with the size distribution shown below:

| Size range number (i) | Size range (μm) | Mass fraction in feed |
|---|---|---|
| 1 | 15–30 | 0.10 |
| 2 | 30–50 | 0.20 |
| 3 | 50–70 | 0.30 |
| 4 | 70–100 | 0.40 |

The bed is fluidized by a gas of density 1.2 kg/m³ and viscosity $18.4 \times 10^{-5}$ Pa s at a superficial gas velocity of 0.4 m/s.
The fluid bed vessel has a cross-sectional area of 1 m².
   Using a discrete time interval calculation with a time increment of 5 min, calculate:

(a)  the size distribution of the bed after 50 min;

(b)  the total mass of solids lost from the bed in that time;

(c)  the maximum solids loading at the process exit;

(d)  the entrainment flux above the transport disengagement height of solids in size range 1 (15–30 μm) after 50 min.

   Assume that the process exit is positioned above TDH and that none of the entrained solids are returned to the bed.

(Answer: (a) {range 1} 0.029, {2} 0.165, {3} 0.324, {4} 0.482; (b) 527 kg; (c) 0.514 kg/ m³; (d) 0.024 kg/m² s)

**5.11** A powder having a particle density of 1800 kg/m³ and the following size distribution:

| Size range number (i) | Size range (μm) | Mass fraction in feed |
|---|---|---|
| 1 | 20–40 | 0.10 |
| 2 | 40–60 | 0.35 |
| 3 | 60–80 | 0.40 |
| 4 | 80–100 | 0.15 |

is fed into a fluidized bed 2 m in diameter at a rate of 0.2 kg/s. The cyclone inlet is 4 m above the distributor and the mass of solids in the bed is held constant at 4000 kg by withdrawing solids continuously from the bed. The bed is fluidized using dry air at 700 K (density 0.504 kg/m$^3$ and viscosity $3.33 \times 10^{-5}$ Pa s) giving a superficial gas velocity of 0.3 m/s. Under these conditions the mean bed voidage is 0.55 and the mean bubble size at the bed surface is 5 cm. For this powder, under these conditions, $U_{mb} = 0.155$ cm/s.

Assuming that none of the entrained solids are returned to the bed, estimate

(a)   the flowrate and size distribution of the entrained solids entering the cyclone;

(b)   the equilibrium size distribution of solids in the bed;

(c)   the solids loading of the gas entering the cyclone;

(d)   the rate at which solids are withdrawn from the bed.

(Answer: (a) 0.0479 kg/s, {range 1} 0.215, {2} 0.420, {3} 0.295, {4} 0.074; (b) {range 1} 0.064, {2} 0.328, {3} 0.433, {4} 0.174; (c) 50 g/m$^3$ (d) 0.152 kg/s)

**5.12** A gas phase catalytic reaction is performed in a fluidized bed operating at a superficial gas velocity equivalent to $10 \times U_{mf}$. For this reaction under these conditions it is known that the reaction is first order in reactant A. Given the following information:

$$kH_{mf}(1 - \varepsilon_p)/U = 100; \chi = \frac{K_C H}{U_B} = 1.0$$

use the reactor model of Orcutt *et al.* to determine:

(a)   the conversion of reactant A,

(b)   the effect on the conversion found in (a) of doubling the inventory of catalyst

(c)   the effect on the conversion found in part (a) of halving the bubble size by using suitable baffles (assuming the interphase mass transfer coefficient is inversely proportional to the bubble diameter)

If the reaction rate were two orders of magnitude smaller, comment on the wisdom of installing baffles in the bed with a view to improving conversion.

(Answer: (a) 0.6645; (b) 0.8744; (c) 0.8706)

# 6

# Pneumatic transport and standpipes

In this chapter we deal with two examples of the transport of particulate solids in the presence of a gas. The first example is pneumatic transport (sometimes referred to as pneumatic conveying), which is the use of a gas to transport a particulate solid through a pipeline. The second example is the standpipe, which has been used for many years, particularly in the oil industry, for transferring solids downwards from a vessel at low pressure to a vessel at a higher pressure.

## 6.1 PNEUMATIC TRANSPORT

For many years gases have been used successfully in industry to transport a wide range of particulate solids – from wheat flour to wheat grain and plastic chips to coal. Until quite recently most pneumatic transport was done in dilute suspension using large volumes of air at high velocity. Since the mid-1960s, however, there has been increasing interest in the so-called 'dense phase' mode of transport in which the solid particles are not fully suspended. The attractions of dense phase transport lie in its low air requirements. Thus, in dense phase transport, a minimum amount of air is delivered to the process with the solids (a particular attraction in feeding solids into fluidized bed reactors, for example). A low air requirement also generally means a lower energy requirement (in spite of the higher pressures needed). The resulting low solids velocities mean that in dense phase transport product degradation by attrition, and pipeline erosion are not the major problems they are in dilute phase pneumatic transport.

In this section we will look at the distinguishing characteristics of dense and dilute phase transport and the types of equipment and systems used with each. The design of dilute phase systems is dealt with in detail and the approach to design of dense phase systems is summarized.

### 6.1.1   Dilute Phase and Dense Phase Transport

The pneumatic transport of particulate solids is broadly classified into two flow regimes: dilute (or lean) phase flow and dense phase flow. Dilute phase flow in its most recognizable form is characterized by high gas velocities (greater than 20 m/s), low solids concentrations (less than 1% by volume) and low pressure drops per unit length of transport line (typically less than 5 mbar/m). Dilute phase pneumatic transport is limited to short route, continuous transport of solids at rates of less than 10 tonnes/hour and is the only system capable of operation under negative pressure. Under these dilute flow conditions the solid particles behave as individuals, fully suspended in the gas, and fluid-particle forces dominate. At the opposite end of the scale is dense phase flow, characterized by low gas velocities (1–5 m/s), high solids concentrations (greater than 30% by volume) and high pressure drops per unit length of pipe (typically greater than 20 mbar/m). In dense phase transport particles are not fully suspended and there is much interaction between the particles.

The boundary between dilute phase flow and dense phase flow, however, is not clear cut and there are as yet no universally accepted definitions of dense phase and dilute phase transport.

Konrad (1986) lists four alternative means of distinguishing dense phase flow from dilute phase flow:

(a)   on the basis of solids/air mass flowrates;

(b)   on the basis of solids concentration;

(c)   dense phase flow exists where the solids completely fill the cross section of the pipe at some point;

(d)   dense phase flow exists when, for horizontal flow, the gas velocity is insufficient to support all particles in suspension, and, for vertical flow, where reverse flow of solids occurs.

In all these cases different authors claim different values and apply different interpretations.

In this chapter the 'choking' and 'saltation' velocities will be used to mark the boundaries between dilute phase transport and dense phase transport in vertical and horizontal pipelines respectively. These terms are defined below in considering the relationships between gas velocity, solids mass flowrate and pressure drop per unit length of transport line in both horizontal and vertical transport.

### 6.1.2   The Choking Velocity in Vertical Transport

We will see in Section 6.1.4 that the pressure drop across a length of transport line has in general six components:

- pressure drop due to gas acceleration;
- pressure drop due to particle acceleration;
- pressure drop due to gas-to-pipe friction;
- pressure drop related to solid-to-pipe friction;
- pressure drop due to the static head of the solids;
- pressure drop due to the static head of the gas.

The general relationship between gas velocity and pressure gradient $\Delta p/\Delta L$ for a vertical transport line is shown in Figure 6.1. Line AB represents the frictional pressure loss due to gas only in a vertical transport line. Curve CDE is for a solids flux of $G_1$ and curve FG is for a higher feed rate $G_2$. At point C the gas velocity is high, the concentration is low, and frictional resistance between gas and pipe wall predominates. As the gas velocity is decreased the frictional resistance decreases but, since the concentration of the suspension increases, the static head required to support these solids increases. If the gas velocity is decreased below point D then the increase in static head outweighs the decrease in frictional resistance and $\Delta p/\Delta L$ rises again. In the region DE the decreasing velocity causes a rapid increase in solids concentration and a point is reached when the gas can no longer entrain all the solids. At this point a flowing, slugging fluidized bed (see Chapter 5) is formed in the transport line. The phenomenon is known as 'choking' and is usually attended by large pressure fluctuations. The choking velocity, $U_{CH}$, is the lowest velocity at which this dilute phase transport line can be operated at the solids feed rate $G_1$. At the higher solids feed rate, $G_2$, the choking velocity is higher. The

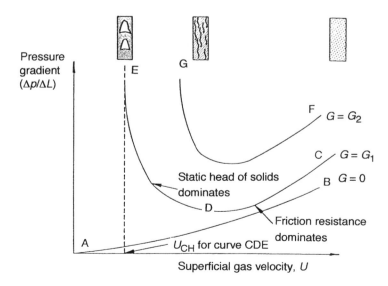

**Figure 6.1**  Phase diagram for dilute-phase vertical pneumatic transport

choking velocity marks the boundary between dilute phase and dense phase vertical pneumatic transport. Note that choking can be reached by decreasing the gas velocity at a constant solids flow rate, or by increasing the solids flow rate at a constant gas velocity.

It is not possible to theoretically predict the conditions for choking to occur. However, many correlations for predicting choking velocities are available in the literature. Knowlton (1986) recommends the correlation of Punwani *et al.* (1976), which takes account of the considerable effect of gas density. This correlation is presented below:

$$\frac{U_{CH}}{\varepsilon_{CH}} - U_T = \frac{G}{\rho_p(1 - \varepsilon_{CH})} \qquad (6.1)$$

$$\rho_f^{0.77} = \frac{2250D(\varepsilon_{CH}^{-4.7} - 1)}{\left[\frac{U_{CH}}{\varepsilon_{CH}} - U_T\right]^2} \qquad (6.2)$$

where $\varepsilon_{CH}$ is the voidage in the pipe at the choking velocity $U_{CH}$, $\rho_p$ is the particle density, $\rho_f$ is the gas density, $G$ is the mass flux of solids ($= M_p/A$) and $U_T$ is the free fall, or terminal velocity, of a single particle in the gas. (Note that the constant is dimensional and that SI units must be used).

Equation (6.1) represents the solids velocity at choking and includes the assumption that the slip velocity $U_{SLIP}$ is equal to $U_T$ (see Section 6.1.4 below for definition of slip velocity). Equations (6.1) and (6.2) must be solved simultaneously by trial and error to give $\varepsilon_{CH}$ and $U_{CH}$.

### 6.1.3   The Saltation Velocity in Horizontal Transport

The general relationship between gas velocity and pressure gradient $\Delta p/\Delta L$ for a horizontal transport line is shown in Figure 6.2 and is in many ways

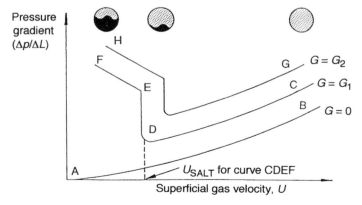

**Figure 6.2**   Phase diagram for dilute-phase horizontal pneumatic transport

similar to that for a vertical transport line. Line AB represents the curve obtained for gas only in the line, CDEF for a solids flux, $G_1$, and curve GH for a higher solids feed rate, $G_2$. At point C, the gas velocity is sufficiently high to carry all the solids in very dilute suspension. The solid particles are prevented from settling to the walls of the pipe by the turbulent eddies generated in the flowing gas. If the gas velocity is reduced whilst solids feed rate is kept constant, the frictional resistance and $\Delta p/\Delta L$ decrease. The solids move more slowly and the solids concentration increases. At point D the gas velocity is insufficient to maintain the solids in suspension and the solids begin to settle out in the bottom of the pipe. The gas velocity at which this occurs is termed the 'saltation velocity'. Further decrease in gas velocity results in rapid 'salting out' of solids and rapid increase in $\Delta p/\Delta L$ as the area available for flow of gas is restricted by settled solids.

In the region $E$ and $F$ some solids may move in dense phase flow along the bottom of the pipe whilst others travel in dilute phase flow in the gas in the upper part of the pipe. The saltation velocity marks the boundary between dilute phase flow and dense phase flow in horizontal pneumatic transport.

Once again, it is not possible to theoretically predict the conditions under which saltation will occur. However, many correlations for predicting saltation velocity are available in the literature. The correlation by Zenz (1964) is frequently used but is entirely empirical and requires the use of a graph. It is reported by Leung and Jones (1978) to have an average error of ±54%. The correlation of Rizk (1973), based on a semi-theoretical approach, is considerably simpler to use, and has a similar error range. It is most unambiguously expressed as

$$\frac{M_p}{\rho_f U_{SALT} A} = \left\{ \frac{1}{10^{(1440x+1.96)}} \right\} \left\{ \frac{U_{SALT}}{\sqrt{gD}} \right\}^{(1100x+2.5)} \tag{6.3}$$

where $\dfrac{M_p}{\rho_f U_{SALT} A}$ is the solids loading $\left[ \dfrac{\text{mass flowrate of solids}}{\text{mass flowrate of gas}} \right]$

$\dfrac{U_{SALT}}{\sqrt{gD}}$ is the Froude Number at saltation

$U_{SALT}$ is the superficial gas velocity (see Section 6.1.4 below for the definition of superficial velocity) at saltation when the mass flowrate of solids is $M_p$, the pipe diameter is D and the particle size is x. (The units are SI).

## 6.1.4   Fundamentals

In this section we generate some basic relationships governing the flow of gas and particles in a pipe.

*Gas and particle velocities*

We have to be careful in the definition of gas and particle velocities and in the relative velocity between them, the slip velocity. The terms are often used loosely in the literature and are defined below.

The term 'superficial velocity' is also commonly used. Superficial gas and solids (particles) velocities are defined as

$$\text{superficial gas velocity, } U_{fs} = \frac{\text{volume flow of gas}}{\text{cross sectional area of pipe}} = \frac{Q_f}{A} \quad (6.4)$$

$$\text{superficial solids velocity, } U_{ps} = \frac{\text{volume flow of solids}}{\text{cross sectional area of pipe}} = \frac{Q_p}{A} \quad (6.5)$$

where subscript 's' denotes superficial and subscripts 'f' and 'p' refer to the fluid and particles respectively.

The fraction of pipe cross-sectional area available for the flow of gas is usually assumed to be equal to the volume fraction occupied by gas, ie. the voidage or void fraction $\varepsilon$. The fraction of pipe area available for the flow of solids is therefore $(1 - \varepsilon)$.

And so, actual gas velocity,

$$U_f = \frac{Q_f}{A\varepsilon} \quad (6.6)$$

and actual particle velocity,

$$U_p = \frac{Q_p}{A(1 - \varepsilon)} \quad (6.7)$$

Thus superficial velocities are related to actual velocities by the equations

$$U_f = \frac{U_{fs}}{\varepsilon} \quad (6.8)$$

$$U_p = \frac{U_{ps}}{1 - \varepsilon} \quad (6.9)$$

It is common practice in dealing with fluidization and pneumatic transport to simply use the symbol $U$ to denote superficial fluid velocity. This practice will be followed in this chapter. Also, in line with common practice, the symbol $G$ will be used to denote the mass flux of solids, i.e. $G = M_p/A$, where $M_p$ is the mass flow rate of solids.

The relative velocity between particle and fluid $U_{rel}$ is defined as

$$U_{rel} = U_f - U_p \quad (6.10)$$

This velocity is often also referred to as the 'slip velocity' $U_{slip}$.

It is often assumed that in vertical dilute phase flow the slip velocity is equal to the single particle terminal velocity $U_T$.

## Continuity

Consider a length of transport pipe into which are fed particles and gas at mass flowrates of $M_p$ and $M_f$ respectively. The continuity equations for particles and gas are:

for the particles:

$$M_p = AU_p(1 - \varepsilon)\rho_p \tag{6.11}$$

for the gas:

$$M_f = AU_f \varepsilon \rho_f \tag{6.12}$$

Combining these continuity equations gives an expression for the ratio of mass flow rates. This ratio is known as the solids loading:

$$\text{Solids loading,} \ \frac{M_p}{M_f} = \frac{U_p(1 - \varepsilon)\rho_p}{U_f \varepsilon \rho_f} \tag{6.13}$$

This shows us that the average voidage $\varepsilon$, at a particular position along the length of the pipe, is a function of the solids loading and the magnitudes of the gas and solids velocities for given gas and particle density.

## Pressure drop

In order to obtain an expression for the total pressure drop along a section of transport line we will write down the momentum equation for a section of pipe. Consider a section of pipe of cross-sectional area $A$ and length $\delta L$ inclined to the horizontal at an angle $\theta$ and carrying a suspension of voidage $\varepsilon$ (see Figure 6.3).

The momentum balance equation is:

$$\begin{bmatrix} \text{net force acting} \\ \text{on pipe contents} \end{bmatrix} = \begin{bmatrix} \text{rate of increase in} \\ \text{momentum of contents} \end{bmatrix}$$

Therefore,

$$\begin{bmatrix} \text{pressure} \\ \text{force} \end{bmatrix} - \begin{bmatrix} \text{gas/wall} \\ \text{friction force} \end{bmatrix} - \begin{bmatrix} \text{solids/wall} \\ \text{friction force} \end{bmatrix} - \begin{bmatrix} \text{gravitational} \\ \text{force} \end{bmatrix}$$

$$= \begin{bmatrix} \text{rate of increase} \\ \text{in momentum} \\ \text{of the gas} \end{bmatrix} + \begin{bmatrix} \text{rate of increase} \\ \text{in momentum} \\ \text{of the solids} \end{bmatrix}$$

or

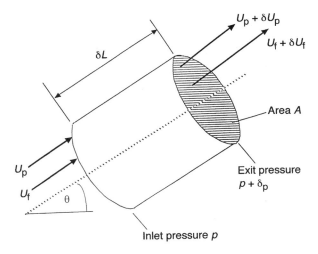

**Figure 6.3**  Section of conveying pipe: basis for momentum equation

$$-A\,\delta p - F_{\mathrm{fw}}A\,\delta L - F_{\mathrm{pw}}A\,\delta L - (A[1-\varepsilon]\rho_{\mathrm{p}}\delta L)g\sin\theta - (A\varepsilon\rho_{\mathrm{f}}\,\delta L)g\sin\theta$$
$$= \rho_{\mathrm{f}}A\varepsilon U_{\mathrm{f}}\,\delta U_{\mathrm{f}} + \rho_{\mathrm{p}}A(1-\varepsilon)U_{\mathrm{p}}\delta U_{\mathrm{p}}\quad(6.14)$$

where $F_{\mathrm{fw}}$ and $F_{\mathrm{pw}}$ are the gas-to-wall friction force and solids-to-wall friction force per unit volume of pipe, respectively.

Rearranging Equation 6.14 and integrating assuming constant gas density and voidage

$$p_1 - p_2 = \underset{(1)}{\tfrac{1}{2}\varepsilon\rho_{\mathrm{f}}U_{\mathrm{f}}^2} + \underset{(2)}{\tfrac{1}{2}(1-\varepsilon)\rho_{\mathrm{p}}U_{\mathrm{p}}^2} + \underset{(3)}{F_{\mathrm{fw}}L} + \underset{(4)}{F_{\mathrm{pw}}L}$$

$$+ \underset{(5)}{\rho_{\mathrm{p}}L(1-\varepsilon)g\sin\theta} + \underset{(6)}{\rho_{\mathrm{f}}L\varepsilon g\sin\theta}\qquad(6.15)$$

Readers should note that Equations (6.4)–(6.15) apply in general to the flow of any gas–particle mixture in a pipe. No assumption has been made as to whether the particles are transported in dilute phase or dense phase.

Equation (6.15) indicates that the total pressure drop along a straight length of pipe carrying solids in dilute phase transport is made up of a number of terms:

(1)  pressure drop due to gas acceleration;

(2)  pressure drop due to particle acceleration;

(3)  pressure drop due to gas-to-wall friction;

(4)  pressure drop related to solid-to-wall friction;

(5)  pressure drop due to the static head of the solids;

(6)  pressure drop due to the static head of the gas.

Some of these terms may be ignored depending on circumstance. If the gas and the solids are already accelerated in the line, then the first two terms should be omitted from the calculation of the pressure drop; if the pipe is horizontal, terms (5) and (6) can be omitted. The main difficulties are in knowing what the solids-to-wall friction is, and whether the gas-to-wall friction can be assumed independent of the presence of the solids; these will be covered in Section 6.1.5.

## 6.1.5   Design for Dilute Phase Transport

Design of a dilute phase transport system involves selection of a combination of pipe size and gas velocity to ensure dilute flow, calculation of the resulting pipeline pressure drop and selection of appropriate equipment for moving the gas and separating the solids from the gas at the end of the line.

### Gas velocity

In both horizontal and vertical dilute phase transport it is desirable to operate at the lowest possible velocity in order to minimize frictional pressure loss, reduce attrition and running costs. For a particular pipe size and solids flowrate, the saltation velocity is always higher than the choking velocity. Therefore, in a transport system comprising both vertical and horizontal lines, the gas velocity must be selected to avoid saltation. In this way choking will also be avoided. These systems would ideally operate at a gas velocity slightly to the right of point D in Figure 6.2. In practice, however, $U_{SALT}$ is not known with great confidence and so conservative design leads to operation well to the right of point D with the consequent increase in frictional losses. Another factor encouraging caution in selecting the design velocity is the fact that the region near to point D is unstable; slight perturbations in the system may bring about saltation.

If the system consists only of a lift line, then the choking velocity becomes the important criterion. Here again, since $U_{CH}$ cannot be predicted with confidence, conservative design is necessary. In systems using a centrifugal blower, characterized by reduced capacity at increased pressure, choking can almost be self-induced. For example, if a small perturbation in the system gives rise to an increase in solids feed rate, the pressure gradient in the vertical line will increase (Figure 6.1). This results in a higher back pressure at the blower giving rise to reduced volume flow of gas. Less gas means higher pressure gradient and the system soon reaches the condition of choking. The system fills with solids and can only be restarted by draining off the solids.

Bearing in mind the uncertainty in the correlations for predicting choking and saltation velocities, safety margins of 50% and greater are recommended when selecting the operating gas velocity.

*Pipeline pressure drop*

Equation (6.15) applies in general to the flow of any gas–particle mixture in a pipe. In order to make the equation specific to dilute phase transport, we must find expressions for terms 3 (gas-to-wall friction) and 4 (solids-to-wall friction).

In dilute transport the gas-to-wall friction is often assumed independent of the presence of the solids and so the friction factor for the gas may be used (e.g. Fanning friction factor – see worked example on dilute pneumatic transport).

Several approaches to estimating solids-to-wall friction are presented in the literature. Here we will use the modified Konno and Saito (1969) correlation for estimating the pressure loss due to solid-to-pipe friction in vertical transport and the Hinkle (1953) correlation for estimating this pressure loss in horizontal transport. Thus for vertical transport (Konno and Saito, 1969):

$$F_{pw}L = 0.057GL\sqrt{\frac{g}{D}} \tag{6.16}$$

and for horizontal transport:

$$F_{pw}L = \frac{2f_P(1 - \varepsilon)\rho_P U_P^2 L}{D} \tag{6.17a}$$

or

$$F_{pw}L = \frac{2f_P G U_P L}{D} \tag{6.17b}$$

where

$$U_p = U(1 - 0.0638\, x^{0.3}\rho_p^{0.5}) \tag{6.18}$$

and (Hinkle, 1953)

$$f_P = \frac{3}{8}\frac{\rho_f}{\rho_p}\frac{D}{x}C_D\left[\frac{U_f - U_p}{U_p}\right]^2 \tag{6.19}$$

where $C_D$ is the drag coefficient between the particle and gas (See Chapter 1).

---

Note:
Hinkle's analysis assumes that particles lose momentum by collision with the pipe walls. The pressure loss due to solids– wall friction is the gas pressure loss as a result of re-accelerating the solids. Thus, from Chapter 1, the drag force on a single particle is given by

$$F_D = \frac{\pi x^2}{4} \rho_f C_D \frac{(U_f - U_p)^2}{2} \tag{6.20}$$

If the void fraction is $\varepsilon$, then the number of particles per unit volume of pipe $N_v$ is

$$N_v = \frac{(1 - \varepsilon)}{\pi x^3 / 6} \tag{6.21}$$

Therefore the force exerted by the gas on the particles in unit volume of pipe $F_v$ is

$$F_v = F_D \frac{(1 - \varepsilon)}{\pi x^3 / 6} \tag{6.22}$$

Based on Hinkle's assumption, this is equal to the solids–wall friction force per unit volume of pipe, $F_{pw}$. Hence,

$$F_{pw} L = \tfrac{3}{4} \rho_f C_D \frac{L}{x} (1 - \varepsilon)(U_f - U_p)^2 \tag{6.23}$$

Expressing this in terms of a friction factor, $f_p$ we obtain Equations (6.17) and (6.19).

Equation (6.15) relates to pressure losses along lengths of straight pipe. Pressure losses are also associated with bends in pipelines and estimations of the value of these losses will be covered in the next section.

*Bends*

Bends complicate the design of pneumatic dilute phase transport systems and when designing a transport system it is best to use as few bends as possible. Bends increase the pressure drop in a line, and also are the points of most serious erosion and particle attrition.

Solids normally in suspension in straight, horizontal or vertical pipes tend to salt out at bends due to the centrifugal force encountered while travelling around the bend. Because of this operation, the particles slow down and are then re-entrained and re-accelerated after they pass through the bend, resulting in the higher pressure drops associated with bends.

There is a greater tendency for particles to salt out in a horizontal pipe which is preceded by a downflowing vertical to horizontal bend than in any other configuration. If this type of bend is present in a system, it is possible for solids to remain on the bottom of the pipe for very long distances following the bend before they redisperse. Therefore, it is recommended that downflowing vertical to horizontal bends be avoided if at all possible in dilute phase pneumatic transport systems.

In the past, designers of dilute phase pneumatic transport systems intuitively thought that gradually sloped, long radius elbows would reduce the erosion and increase bend service life relative to 90° elbows. Zenz (1964), however, recommended that blinded tees (Figure 6.4) be used in place of elbows in pneumatic transport systems. The theory behind the use of the blinded tee is that a cushion of stagnant particles collects in the blinded or unused branch of the tee, and the conveyed particles then impinge upon the stagnant particles in the tee rather than on the metal surface, as in a long radius or short radius elbow. Bodner (1982) determined the service life and pressure drop of various bend configurations. He found that the service life of the blinded tee configuration was far better than any other configuration tested and that it gave a service life 15 times greater than that of radius bends or elbows. This was due to the cushioning accumulation of particles in the blinded branch of the tee which he observed in glass bend models. Bodner also reported that pressure drops and solid attrition rates for the blinded tee were approximately the same as those observed for radius bends.

In spite of a considerable amount of research into bend pressure drop, at present there is no reliable method of predicting accurate bend pressure drops other than by experiment for the actual conditions expected. In industrial practice, bend pressure drop is often approximated by assuming that it is equivalent to approximately 7.5 m of vertical section pressure drop. In the absence of any reliable correlation to predict bend pressure drop, this crude method is probably as reliable and as conservative as any.

## Equipment

Dilute phase transport is carried in systems in which the solids are fed into the air stream. Solids are fed from a hopper at a controlled rate through a rotary air lock into the air stream. The system may be positive pressure, negative pressure or employ a combination of both. Positive pressure systems are

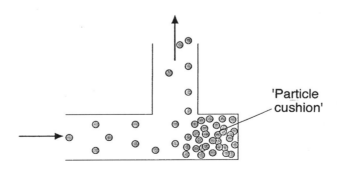

**Figure 6.4**   Blinded tee bend

usually limited to a maximum pressure of 1 bar gauge and negative pressure systems to a vacuum of about 0.4 bar by the types of blowers and exhausters used.

Typical dilute phase systems are shown in Figures 6.5 and 6.6. Blowers are normally of the positive displacement type which may or may not have speed control in order to vary volume flowrate. Rotary airlocks enable solids to be fed at a controlled rate into the air stream against the air pressure. Screw

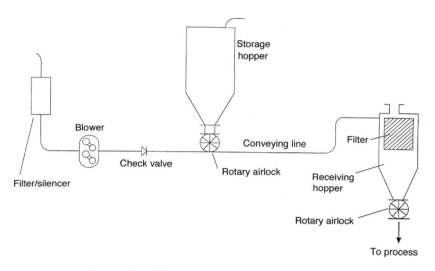

**Figure 6.5**    Dilute-phase transport: positive pressure system

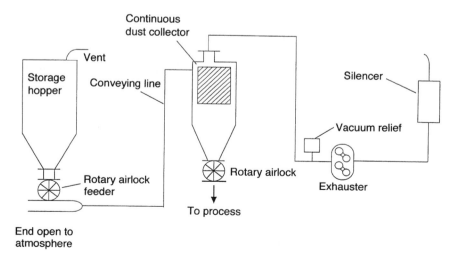

**Figure 6.6**    Dilute-phase transport: negative pressure system

feeders are frequently used to transfer solids. Cyclone separators (see Chapter 7) are used to recover the solids from the gas stream at the receiving end of the transport line. Filters of various types and with various methods of solids recovery are used to clean up the transport gas before discharge or recycle.

In some circumstances it may not be desirable to use once-through air as the transport gas (e.g. for the risk of contamination of the factory with toxic or radioactive substances; for risk of explosion an inert gas may be used; in order to control humidity when the solids are moisture sensitive). In these cases a closed loop system is used. If a rotary positive displacement blower is used then the solids must be separated from the gas by cyclone separator and by in-line fabric filter. If lower system pressures are acceptable (0.2 bar gauge) then a centrifugal blower may be used in conjunction with only a cyclone separator. The centrifugal fan is able to pass small quantities of solids without damage, whereas the positive displacement blower will not pass dust.

### 6.1.6   Dense Phase Transport

*Flow patterns*

As pointed out in the introduction to this chapter, there are many different definitions of dense phase transport and of the transition point between dilute phase and dense phase transport. For the purpose of this section dense phase transport is described as the condition in which solids are conveyed such that they are not entirely suspended in the gas. Thus, the transition point between dilute and dense phase transport is saltation for horizontal transport and choking for vertical transport.

However, even within the dense phase regime a number of different flow patterns occur in both horizontal and vertical transport. Each of these flow patterns has particular characteristics giving rise to particular relationships between gas velocity, solids flowrate and pipeline pressure drop. In Figure 6.7 for example, five different flow patterns are identified within the dense phase regime for horizontal transport.

The continuous dense phase in which the solids occupy the entire pipe is virtually extrusion. Transport in this form requires very high gas pressures and is limited to short straight pipe lengths and granular materials (which have a high permeability).

Discontinuous dense phase flow can be divided into three fairly distinct flow patterns: 'discrete plug flow' in which discrete plugs of solids occupy the full pipe cross section; 'dune flow' in which a layer of solids settled at the bottom of the pipe move along in the form of rolling dunes; a hybrid of discrete plug flow and dune flow in which the rolling dunes completely fill the pipe cross-section but in which there are no discrete plugs (also known as 'plug flow').

Saltating flow is encountered at gas velocities just below the saltation velocity. Particles are conveyed in suspension above a layer of settled solids. Particles may be deposited and re-entrained from this layer. As the gas

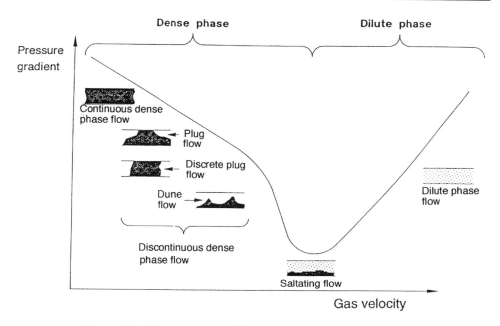

**Figure 6.7**    Flow patterns in horizontal pneumatic conveying

velocity is decreased the thickness of the layer of settled solids increases and eventually we have dune flow.

It should be noted, firstly, that not all powders exhibit all these flow patterns and, secondly, that within any transport line it is possible to encounter more than one regime.

The main advantages of dense phase transport arise from the low gas requirements and low solids velocities. Low gas volume requirements generally mean low energy requirements per kilogram of product conveyed, and also mean that smaller pipelines and recovery and solids–gas separation are required. Indeed in some cases, since the solids are not suspended in the transport gas, it may be possible to operate without a filter at the receiving end of the pipeline. Low solids velocities means that abrasive and friable materials may be conveyed without major pipeline erosion or product degradation.

It is interesting to look at the characteristics of the different dense phase flow patterns with a view to selecting the optimum for a dense phase transport system. The continuous dense phase flow pattern is the most attractive from the point of view of low gas requirements and solid velocities, but has the serious drawback that it is limited to use in the transport of granular materials along short straight pipes and requires very high pressures. Saltating flow occurs at a velocity too close to the saltation velocity and is therefore unstable. In addition this flow pattern offers little advantage in the area of gas and solids velocity. We are then left with the so-called discontinuous dense phase flow pattern with its plugs and dunes. However, performance in this area is unpredictable, can give rise to complete pipeline blockages and requires high pressures. Most commercial dense phase transport systems operate in this flow pattern and incorporate some means of controlling plug length in order to increase predictability and reduce the chance of blockages.

It is therefore necessary to consider how the pressure drop across a plug of solids depends on its length. Unfortunately contradictory experimental evidence is reported in the literature. Konrad (1986) points out that the pressure drop across a moving plug has been reported to increase (a) linearly with plug length, (b) as the square of the plug length and (c) exponentially with plug length. A possible explanation of these apparent contradictions is reported by Klintworth and Marcus (1985) who cite the work of Wilson (1981) on the effect of stress on the deformation within the plug. Large cohesionless particles (typically Geldart Group D particles {Geldart's classification of powders – see Chapter 5 on Fluidization}) give rise to a permeable plug permitting the passage of a significant gas flow at low pressure drops. In this case the stress developed in the plug would be low and a linear dependence of pressure drop on plug length would result. Plugs of fine cohesive particles (typically Geldart Group C) would be virtually impermeable to gas flow at the pressures usually encountered. In this case, the plug moves as a piston in a cylinder by purely mechanical means. The stress developed within the plug is high. The high stress translates to a high wall shear stress which gives rise to an exponential increase in pressure drop with plug length. Thus it is the degree of permeability of the plug which determines the relationship between plug length and pressure drop. The pressure drop across a plug can vary between a linear and an exponential function of the plug length depending on the permeability of the plug.

Large cohesionless particles form permeable plugs and are therefore suitable for discontinuous dense phase transport. In other materials, where interaction under the action of stress and interparticle forces give rise to low permeability plugs, discontinuous dense phase transport is only possible if some mechanism is used to limit plug length, avoiding blockages.

*Equipment*

In commercial systems, the problem of plug formation is tackled in three ways:

1. Detect the plug at its formation and take appropriate action to either

   (a) use a bypass system in which the pressure build-up behind a plug causes more air to flow around the bypass line and break up the plug from its front end (Figure 6.8);

   (b) detect the pressure build-up using pressure actuated valves which divert auxiliary air to break up the plugs into smaller lengths, (Figure 6.9).

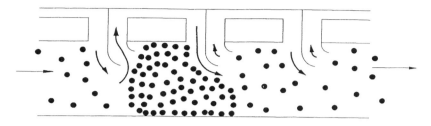

**Figure 6.8**  Dense phase conveying system using a bypass line to break up plugs of solids

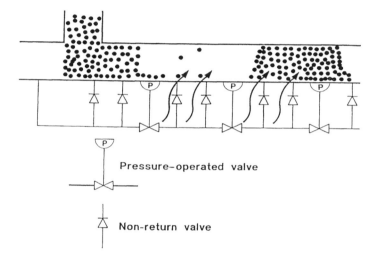

**Figure 6.9**  Dense phase conveying system using pressure-actuated valves to direct gas

2.  Form stable plugs – stable plugs of granular material do form naturally under certain conditions. However, to form stable plugs of manageable length of other materials, it is generally necessary to induce them artificially by one of the following means:

    (a)  use an air knife to chop up solids fed in continuous dense phase flow from a blow tank (Figure 6.10);

    (b)  use an alternating valves system (Figure 6.11) in order to cut up the continuous dense phase flow from the blow tank;

    (c)  for free-flowing materials it is possible to use an air operated diaphragm in the blow tank to create plugs (Figure 6.12);

    (d)  a novel idea reported by Tsuji (1983) uses table tennis balls to separate solids into plugs.

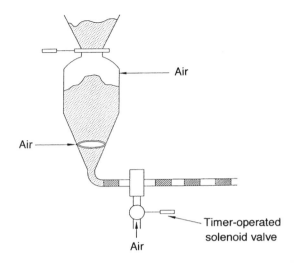

**Figure 6.10**   Solid plug formation using timer-operated air knife

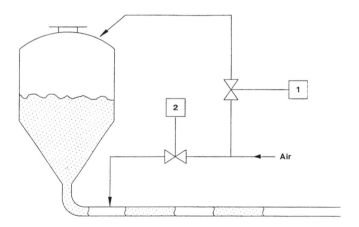

**Figure 6.11**   Solid plug formation using alternating air valves (valves 1 and 2 open and close alternately to create plugs of solids in the discharge pipe

3.   Fluidization – add extra air along the transport line in order to maintain the aeration of the solids and hence avoid the formation of blockages.

Whatever the mechanism used to tackle the plug problem, all commercial dense phase transport systems employ a blow-tank which may be with fluidizing element (Figure 6.13) or without (Figure 6.14).

The blow tank is automatically taken through repeated cycles of filling, pressurising and discharging. Since one third of the cycle time is used for filling the blowtank, a system required to give a mean delivery rate of

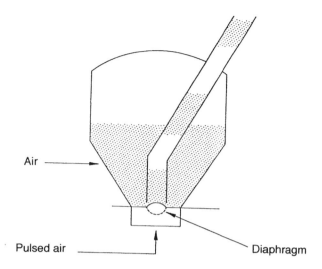

**Figure 6.12** Solid plug creation using air-operated diaphragm

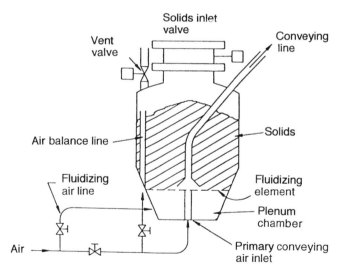

**Figure 6.13** Dense phase transport blow tank with fluidizing element

20 tonnes/hour must be able to deliver a peak rate of over 30 tonnes/hour. Dense phase transport is thus a batch operation because of the high pressures involved, whereas dilute phase transport can be continuous because of the relatively low pressures and the use of rotary valves. The dense phase system can be made to operate in semi-continuous mode by the use of two blow tanks in parallel.

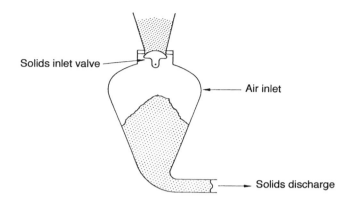

**Figure 6.14**   Blow tank without fluidizing element

## Design for dense phase transport

Whereas dilute phase transport systems can be designed, albeit with a large
safety margin, from first principles together with the help of some empirical
correlations, the design of commercial dense phase systems is largely empirically based. Although in theory the equation for pressure drop in two phase
flow developed earlier in this chapter (Equation (6.15)) may be applied to
dense phase flow, in practice it is of little use. Generally a test facility which
can be made to simulate most transport situations is used to monitor the
important transport parameters during tests on a particular material. From
these results, details of the dense phase transport characteristics of the material
can be built up and the optimum conditions of pipe size, air flowrate, and type
of dense phase system can be determined. Commercial dense phase systems
are designed on the basis of past experience together with the results of tests
such as these. Details of how this is done may be found in Mills (1990).

### 6.1.7   Matching the System to the Powder

Generally speaking it is possible to convey any powder in the dilute phase
mode, but because of the attractions of dense phase transport, there is great
interest in assessing the suitability of a powder for transport in this mode.
The most commonly used procedure is to undertake a series of tests on a
sample of the powder in a pilot plant. This is obviously expensive. An
alternative approach offered by Dixon (1979) is quite widely used. Dixon
recognized the similarities between gas fluidization and dense phase transport and proposed a method of assessing the suitability of a powder for
transport in the dense phase based on Geldart's (1973) classification of
powders (see Chapter 5 on Fluidization). Dixon proposed a 'slugging diagram' which allows prediction of the possible dense phase flow patterns from
a knowledge of particle size and density. Dixon concluded that Geldart's

Groups A and D were suitable for dense phase transport whereas Groups B and C were generally not suitable.

Mainwaring and Reed (1987) claim that although Dixon's approach gives a good general indication of the most likely mode of dense phase transport, it is not the most appropriate means of determining whether a powder will convey in this mode in the first instance. These authors propose an assessment based on the results of bench-scale measurements of the permeability and de-aeration characteristics of the powder. On this basis powders achieving a sufficiently high permeability in the test would be suitable for plug type dense phase transport and powders scoring high on air retention would be suitable for transport in the rolling dune mode of dense phase flow. According to the authors, powders satisfying neither of these criteria are unsuitable for transport by conventional blow tank systems. Flain (1972) offered a qualitative approach to matching the powder to the system. He lists twelve devices for bringing about the initial contact between gas and solids in a transport system and matches powder characteristics to device. This is a useful starting point since certain equipment can be excluded for use with a particular powder.

## 6.2   STANDPIPES

Standpipes have been used for many years, particularly in the oil industry, for transferring solids downwards from a region of low pressure to a region of higher pressure. The overview of standpipe operation given here is based largely on the work of Knowlton (1997).

Typical overflow and underflow standpipes are shown in Figure 6.15, where they are used to continuously transfer solids from an upper fluidized bed to a lower fluidized bed. For solids to be transferred downwards against the pressure gradient gas must flow upward relative to the solids. The friction losses developed by the flow of the gas through the packed or fluidized bed of solids in

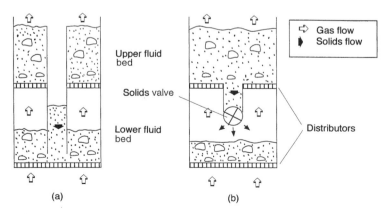

**Figure 6.15**   (a) Overflow and (b) underflow type standpipes transporting solids from low pressure fluidized bed to bed at higher pressure

the standpipe generates the required pressure gradient. If the gas must flow upwards relative to the downflowing solids there are two possible cases: (i) the gas flows upward relative to the standpipe wall and (ii) the gas flows downwards relative to the standpipe wall, but at a lower velocity than the solids.

A standpipe may operate in two basic flow regimes depending on the relative velocity of the gas to the solids; packed bed flow and fluidized bed flow.

### 6.2.1 Standpipes in Packed Flow

If the relative upward velocity of the gas $(U_f - U_p)$ is less than the relative velocity at incipient fluidization $(U_f - U_p)_{mf}$, then packed bed flow results and the relationship between gas velocity and pressure gradient is in general determined by the Ergun equation (see Chapter 4, Equation (4.11)).

The Ergun equation is usually expressed in terms of the superficial gas velocity through the packed bed. However, for the purposes of standpipe calculations it is useful to write the Ergun equation in terms of the magnitude of the velocity of the gas relative to the velocity of the solids $|U_{rel}|\{= |U_f - U_p|\}$. (Refer to Section 6.1.4 for clarification of relationships between superficial and actual velocities.)

$$\text{Superficial gas velocity, } U = \varepsilon|U_{rel}| \tag{6.24}$$

And so in terms of $|U_{rel}|$ the Ergun equation becomes

$$\frac{(-\Delta p)}{H} = \left[150\frac{\mu}{x_{sv}^2}\frac{(1-\varepsilon)^2}{\varepsilon^2}\right]|U_{rel}| + \left[1.75\frac{\rho_f}{x_{sv}}\frac{(1-\varepsilon)}{\varepsilon}\right]|U_{rel}|^2 \tag{6.25}$$

The equation allows us to calculate the value of $|U_{rel}|$ required to give a particular pressure gradient. We now adopt a sign convention for velocities. For standpipes it is convenient to take downward velocities as positive. In order to create the pressure gradient in the required direction (higher pressure at the lower end of the standpipe), the gas must flow upwards relative to the solids. Hence, $U_{rel}$ should always be negative in normal operation. Solids flow is downwards, so $U_p$, the actual velocity of the solids (relative to the pipe wall), is always positive.

Knowing the magnitude and direction of $U_p$ and $U_{rel}$, the magnitude and direction of the actual gas velocity (relative to the pipe wall) may be found from $U_{rel} = U_f - U_p$. In this way the quantity of gas passing up or down the standpipe may be estimated.

### 6.2.2 Standpipes in Fluidized Bed Flow

If the relative upward velocity of the gas $(U_f - U_p)$ is greater than the relative velocity at incipient fluidization $(U_f - U_p)_{mf}$, then fluidized bed flow will result. In fluidized bed flow the pressure gradient is independent of relative gas velocity. Assuming that in fluidized bed flow the entire apparent weight of the particles is supported by the gas flow, then the pressure gradient is given by (see Chapter 5):

$$\frac{(-\Delta p)}{H} = (1 - \varepsilon)(\rho_p - \rho_f)g \qquad (6.26)$$

where $(-\Delta p)$ is the pressure drop across a height $H$ of solids in the standpipe, $\varepsilon$ is the voidage and $\rho_p$ is the particle density.

Fluidized bed flow may be non-bubbling flow or bubbling flow. Non-bubbling flow occurs only with Geldart Group A solids (described in Chapter 5) when the relative gas velocity lies between the relative velocity for incipient fluidization and the relative velocity for minimum bubbling $(U_f - U_p)_{mb}$. For Geldart Group B materials (Chapter 5) with $(U_f - U_p) > (U_f - U_p)_{mf}$ and for Group A solids with $(U_f - U_p) > (U_f - U_p)_{mb}$ bubbling fluidized flow results.

Four types of bubbling fluidized bed flow in standpipes are possible depending on the direction of motion of the gas in the bubble phase and emulsion phases relative to the standpipe walls. These are depicted in Figure 6.16. In

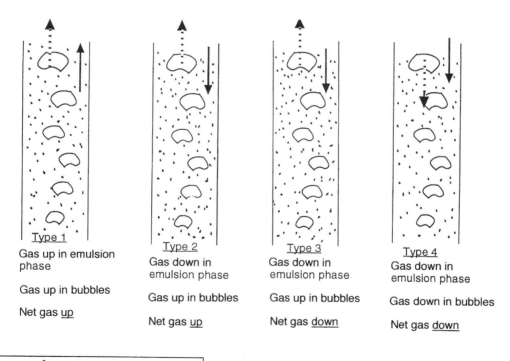

Type 1
Gas up in emulsion phase

Gas up in bubbles

Net gas up

Type 2
Gas down in emulsion phase

Gas up in bubbles

Net gas up

Type 3
Gas down in emulsion phase

Gas up in bubbles

Net gas down

Type 4
Gas down in emulsion phase

Gas down in bubbles

Net gas down

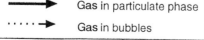

→  Gas in particulate phase

······►  Gas in bubbles

**Figure 6.16**   Types of fluidized flow in a standpipe

practice, bubbles are undesirable in a standpipe. The presence of rising bubbles hinders the flow of solids and reduces the pressure gradient developed in the standpipe. If the bubble rise velocity is greater than the solids velocity, then the bubbles will rise and grow by coalescence. Larger standpipes are easier to operate since they can tolerate larger bubbles than smaller standpipes. For optimum standpipe operation, when using Group B solids, the relative gas velocity should be slightly greater than relative velocity for incipient fluidization. For Group A solids, relative gas velocity should lie between $(U_f - U_p)_{mf}$ and $(U_f - U_p)_{mb}$.

In practice, aeration is often added along the length of a standpipe in order to maintain the solids in a fluidized state just above minimum fluidization velocity. If this were not done then, with a constant mass flow of gas, relative velocities would decrease towards the high-pressure end of the standpipe. The lower velocities would result in lower mean voidages and the possibility of an unfluidized region at the bottom of the standpipe. Aeration gas is added in stages along the length of the standpipe and only the minimum requirement is added at any level. If too much is added, bubbles are created which may hinder solids flow. The analysis below, based on that of Kunii and Levenspiel (1991), enables calculation of the position and quantity of aeration gas to be added.

The starting point is Equation (6.13), the equation derived from the continuity equations for gas and solids flow in a pipe. For fine Group A solids in question the relative velocity between gas and particles will be very small in comparison with the actual velocities, and so we can assume with little error that $U_p = U_f$. Hence, from Equation (6.13):

$$\frac{M_p}{M_f} = \frac{(1 - \varepsilon)}{\varepsilon} \frac{\rho_p}{\rho_f} \tag{6.27}$$

Using subscripts 1 and 2 to refer to the upper (low pressure) and lower (high pressure) level in the standpipe, since $M_p$, $M_f$ and $\rho_p$ are constant:

$$\frac{(1 - \varepsilon_1)}{\varepsilon_1} \frac{1}{\rho_{f_1}} = \frac{(1 - \varepsilon_2)}{\varepsilon_2} \frac{1}{\rho_{f_2}} \tag{6.28}$$

And so, since the pressure ratio $p_2/p_1 = \rho_{f_2}/\rho_{f_1}$, then

$$\frac{p_2}{p_1} = \frac{(1 - \varepsilon_2)}{\varepsilon_2} \frac{\varepsilon_1}{(1 - \varepsilon_1)} \tag{6.29}$$

Let us assume that the voidage $\varepsilon_2$ is the lowest voidage acceptable for maintaining fluidized standpipe flow. Equation (6.29) allows calculation of the equivalent maximum pressure ratio, and hence the pressure drop between levels 1 and 2. Assuming the solids are fully supported, this pressure difference will be equal to the apparent weight per unit cross-sectional area of the standpipe (Equation (6.26)).

$$(p_2 - p_1) = (\rho_p - \rho_f)(1 - \varepsilon_a)Hg \tag{6.30}$$

where $\varepsilon_a$ is the average voidage over the section between levels 1 and 2, $H$ is the distance between the levels and $g$ is the acceleration due to gravity.

If $\varepsilon_1$ and $\varepsilon_2$ are known and gas density is regarded as negligible compared to particle density, $H$ may be calculated from Equation (6.30).

The objective of adding aeration gas is to raise the voidage at the lower level to equal that at the upper level. Applying Equation (6.27),

$$\frac{(1-\varepsilon_2)}{\varepsilon_2} = \frac{M_p}{(M_f + M_{f2})}\frac{\rho_{f2}}{\rho_p} = \frac{M_p}{M_f}\frac{\rho_{f1}}{\rho_p} \tag{6.31}$$

where $M_{f_2}$ is the mass flow of aeration air added at level 2. Then rearranging,

$$M_{f_2} = M_f\left[\frac{\rho_{f_2}}{\rho_{f_1}} - 1\right] \tag{6.32}$$

and, since from Equation 6.27, $M_f = M_p\dfrac{\varepsilon_1}{(1-\varepsilon_1)}\dfrac{\rho_{f_1}}{\rho_p}$

$$M_{f_2} = M_p\frac{\varepsilon_1}{(1-\varepsilon_1)}\frac{\rho_{f_1}}{\rho_p}\left[\frac{\rho_{f_2}}{\rho_{f_1}} - 1\right] \tag{6.33}$$

and so mass flow of aeration air to be added,

$$M_{f_2} = \frac{\varepsilon_1}{(1-\varepsilon_1)}\frac{M_p}{\rho_p}(\rho_{f_2} - \rho_{f_1}) \tag{6.34}$$

from which, it can also be shown that

$$Q_{f_2} = Q_p\frac{\varepsilon_1}{(1-\varepsilon_1)}\left[1 - \frac{\rho_{f_1}}{\rho_{f_2}}\right] \tag{6.35}$$

where $Q_{f2}$ is the volume flow rate of gas to be added at pressure $p_2$ and $Q_p$ is the volume flow rate of solids down the standpipe.

For long standpipes aeration gas will need to be added at several levels in order to keep the voidage within the required range (see the worked example on standpipe aeration at the end of this chapter).

## 6.2.3   Pressure Balance During Standpipe Operation

As an example of the operation of a standpipe, we will consider how an overflow standpipe operating in fluidized bed flow reacts to a change in gas flow rate. Figure 6.17(a) shows the pressure profile over such a system. The pressure balance equation over this system is

$$\Delta p_{SP} = \Delta p_{LB} + \Delta p_d + \Delta p_{UB} \tag{6.36}$$

where $\Delta P_{SP}$, $\Delta P_{LB}$, $\Delta P_{UB}$ and $\Delta P_d$ are the pressure drops across the standpipe, the lower fluidized bed, the upper fluidized bed and the distributor of the upper fluidized bed respectively.

Let us consider a disturbance in the system such that the gas flow through the fluidized beds increases (Figure 6.17(b)). If the gas flow through the lower

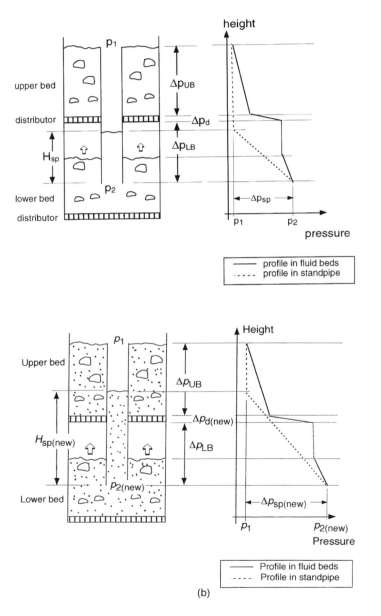

**Figure 6.17** Operation of an overflow standpipe: (a) before increasing gas flow; (b) change in pressure profile due to increased gas flow through the fluid beds

bed increases, although the pressure drops across the lower and upper beds will remain constant, the pressure drop across the upper distributor will increase $\Delta P_{d(new)}$. To match this increase, the pressure across the standpipe must rise to $\Delta P_{SP(new)}$ (Figure 6.17(b)). In the case of an overflow standpipe operating in fluidized flow the increase in standpipe pressure drop results from a rise in the height of solids in the standpipe to $H_{SP(new)}$.

Consider now the case of an underflow standpipe operating in packed bed flow (Figure 6.18), the pressure balance across the system is given by

$$\Delta p_{SP} = \Delta p_v + \Delta p_d \tag{6.37}$$

where $\Delta P_{SP}$, $\Delta P_d$ and $\Delta P_v$ are the pressure drops across the standpipe, the distributor of the upper fluidized bed and the standpipe valve respectively.

If the gas flow from the lower bed increases, the pressure drop across the upper bed distributor increases to $\Delta P_{d(new)}$. The pressure balance then calls for an increase in standpipe pressure drop. Since in this case the standpipe length is fixed, in packed bed flow this increase in pressure drop is achieved by an increase in the magnitude of the relative velocity $|U_{rel}|$. The standpipe pressure drop will increase to $\Delta P_{SP(new)}$ and the valve pressure drop, which depends on the solids flow, will remain essentially constant. Once the standpipe pressure gradient reaches that required for fluidized bed flow, its pressure drop will remain constant so it will not be able to adjust to system changes.

A standpipe commonly used in the petroleum industry is the underflow vertical standpipe with slide valve at the lower end. In this case the standpipe generates more head than is required and the excess is used across the slide valve in controlling the solids flow. Such a standpipe is used in the fluid catalytic cracking (FCC) unit to transfer solids from the reactor to the regenerator.

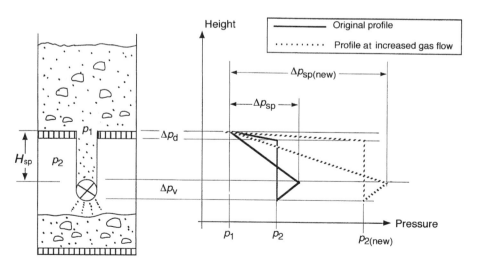

**Figure 6.18** Pressure balance during operation of an underflow standpipe: effect of increasing gas flow through fluid beds

## 6.3   FURTHER READING

Readers wishing to learn more about solids circulation systems, standpipe flow and non-mechanical valves are referred to Kunii and Levenspiel (1991) or the chapters by Knowlton in either Geldart (1986) or Grace *et al.* (1997).

## 6.4   WORKED EXAMPLES

### WORKED EXAMPLE 6.1

Design a positive pressure dilute-phase pneumatic transport system to transport 900 kg/hr of sand of particle density 2500 kg/m³ and mean particle size 100 μm between two points in a plant separated by 10 metres vertical distance and 30 metres horizontal distance using ambient air. Assume that six 90 degree bends are required and that the allowable pressure loss is 0.55 bar.

### Solution

In this case, to design the system means to determine the pipe size and air flowrate which would give a total system pressure loss near to the allowable pressure loss.

The design procedure requires trial and error calculations. Pipes are available in fixed sizes and so the procedure adopted here is to select a pipe size and determine the saltation velocity from Equation (6.1). The system pressure loss is then calculated at a superficial gas velocity equal to 1.5 times the saltation velocity (this gives a reasonable safety margin bearing in mind the accuracy of the correlation in Equation (6.1)). The calculated system pressure loss is then compared with the allowable pressure loss. The pipe size selected may then be altered and the above procedure repeated until the calculated pressure loss matches that allowed.

### Step 1.   Selection of pipe size

Select 78 mm internal diameter pipe.

### Step 2.   Determine gas velocity

Use the Rizk correlation of Equation (6.3) to estimate the saltation velocity, $U_{SALT}$. Equation (6.3) rearranged becomes

$$U_{SALT} = \left[ \frac{4M_p 10^\alpha g^{\beta/2} D^{(\beta/2)-2}}{\pi \rho_f} \right]^{1/(\beta+1)}$$

where $\alpha = 1440x + 1.96$ and $\beta = 1100x + 2.5$.
In the present case $\alpha = 2.104$, $\beta = 2.61$ and $U_{SALT} = 9.88$.
Therefore, superficial gas velocity, $U = 1.5 \times 9.88$ m/s $= 14.82$ m/s.

## Step 3. Pressure loss calculations

(a) *Horizontal sections.* Starting with Equation (6.15) an expression for the total pressure loss in the horizontal sections of the transport line may be generated. We will assume that all the initial acceleration of the solids and the gas take place in the horizontal sections and so terms 1 and 2 are required. For term 3 the Fanning friction equation is used assuming that the pressure loss due to gas/wall friction is independent of the presence of solids. For term 4 we employ the Hinkle correlation (Equation (6.17)). Terms 5 and 6 became zero as $\theta = 0$ for horizontal pipe. Thus, the pressure loss, $\Delta p_H$, in the horizontal sections of the transport line is given by

$$\Delta p_H = \frac{\rho_f \varepsilon_H U_{fH}^2}{2} + \frac{\rho_p(1 - \varepsilon_H)U_{pH}^2}{2} + \frac{2f_g\rho_f U^2 L_H}{D} + \frac{2f_p\rho_p(1 - \varepsilon_H)U_{pH}^2 L_H}{D}$$

where the subscript H refers to the values specific to the horizontal sections.

To use this equation we need to know $\varepsilon_H$, $U_{fH}$ and $U_{pH}$. Hinkle's correlation gives us $U_{pH}$:

$$U_{pH} = U \ (1 - 0.0638x^{0.3} \ \rho_p^{0.5}) = 11.84 \text{ m/s}$$

From continuity, $G = \rho_p(1 - \varepsilon_H)U_{pH}$

$$\text{thus } \varepsilon_H = 1 - \frac{G}{\rho_p U_{pH}} = 0.9982$$

$$\text{and } U_{fH} = \frac{U}{\varepsilon_H} = \frac{14.82}{0.9982} = 14.85 \text{ m/s}$$

Friction factor $f_p$ is found from Equation (6.19) with $C_D$ estimated at the relative velocity $(U_f - U_p)$, using the approximate correlations given below, (or by using an appropriate $C_D$ versus $Re$ chart [see Chapter 1]),

$$\begin{array}{ll} Re_p < 1: & C_D = 24/Re_p \\ 1 < Re_p < 500: & C_D = 18.5 Re_p^{-0.6} \\ 500 < Re_p < 2 \times 10^5: & C_D = 0.44 \end{array}$$

Thus, for flow in the horizontal sections,

$$Re_p = \frac{\rho_f(U_{fH} - U_{pH})x}{\mu}$$

for ambient air $\rho_f = 1.2 \text{ kg/m}^3$ and $\mu = 18.4 \times 10^{-6} \text{ Pa s}$, giving

$$Re_p = 19.63$$

and so, using the approximate correlations above,

$$C_D = 18.5 Re_p^{-0.6} = 3.1$$

Substituting $C_D = 3.1$ in Equation (6.19) we have

$$f_P = \tfrac{3}{8} \times \frac{1.2}{2500} \times 3.1 \times \frac{0.078}{100 \times 10^{-6}} \left\{ \frac{14.85 - 11.84}{11.84} \right\}^2$$

The gas friction factor is taken as $f_g = 0.005$. This gives $\Delta p_H = 14\,864$ Pa.

(b) *Vertical Sections*. Starting again with Equation (6.15), the general pressure loss equation, an expression for the total pressure loss in the vertical section may be derived. Since the initial acceleration of solids and gas was assumed to take place in the horizontal sections, terms 1 and 2 become zero. The Fanning friction equation is used to estimate the pressure loss due to gas-to-wall friction (term 3) assuming solids have negligible effect on this pressure loss. For term 4 the modified Konno and Saito correlation (equation (6.16)) is used. For vertical transport $\theta$ becomes equal to 90° in terms 5 and 6.

   Thus, the pressure loss, $\Delta p_v$, in the vertical sections of the transport line is given by

$$\Delta p_v = \frac{2 f_g \rho_f U^2 L_v}{D} + 0.057 G L_v \sqrt{\frac{g}{D}} + \rho_p (1 - \varepsilon_v) g L_v + \rho_f \varepsilon_v g L_v$$

where subscript v refers to values specific to the vertical sections.

   To use this equation we need to calculate the voidage of the suspension in the vertical pipe line $\varepsilon_v$:

Assuming particles behave as individuals, then slip velocity is equal to single particle terminal velocity, $U_T$ (also noting that the superficial gas velocity in both horizontal and vertical sections is the same and equal to $U$), i.e.

$$U_{pv} = \frac{U}{\varepsilon_v} - U_T$$

continuity gives particle mass flux, $G = \rho_p (1 - \varepsilon_v) U_{pv}$.

Combining these equations gives a quadratic in $\varepsilon_v$ which has only one possible root.

$$\varepsilon_v^2 U_T - \left[ U_T + U + \frac{G}{\rho_p} \right] \varepsilon_v + U = 0$$

The single particle terminal velocity, $U_T$ may be estimated as shown in Chapter 1, giving $U_T = 0.52$ m/s assuming the particles are spherical.

And so, solving the quadratic equation, $\varepsilon_v = 0.9985$
and thus $\Delta p_v = 1148$ Pa

(c) *Bends*. The pressure loss across each 90° bend is taken to be equivalent to that across 7.5 m of vertical pipe.

$$\text{Pressure loss per metre of vertical pipe} = \frac{\Delta p_v}{L_v} = 114.8 \text{ Pa/m}$$

Therefore, pressure loss across six 90° bends

$$= 6 \times 7.5 \times 114.8 \text{ Pa}$$

$$= 5166 \text{ Pa}$$

And so,

$$\begin{bmatrix} \text{total pressure} \\ \text{loss} \end{bmatrix} = \begin{bmatrix} \text{loss across} \\ \text{vertical sections} \end{bmatrix} + \begin{bmatrix} \text{loss across} \\ \text{horizontal} \\ \text{sections} \end{bmatrix} + \begin{bmatrix} \text{loss across} \\ \text{bends} \end{bmatrix}$$

$$= 1148 + 14\,864 + 5166 \text{ Pa}$$
$$= 0.212 \text{ bar}$$

## Step 4. *Compare calculated and allowable pressure losses*

The allowable system pressure loss is 0.55 bar and so we may select a smaller pipe size and repeat the above calculation procedure. The table below gives the results for a range of pipe sizes.

| Pipe inside diameter (mm) | Total System Pressure Loss (bar) |
|---|---|
| 78 | 0.212 |
| 63 | 0.322 |
| 50 | 0.512 |
| 40 | 0.809 |

In this case we would select 50 mm pipe which gives a total system pressure loss of 0.512 bar. (An economic option could be found if capital and running cost were incorporated). The design details for this selection are given below:

$$\text{pipe size} = 50 \text{ mm inside diameter}$$
$$\text{air flowrate} = 0.0317 \text{ m}^3/\text{s}$$
$$\text{air superficial velocity} = 16.15 \text{ m/s}$$
$$\text{saltation velocity} = 10.77 \text{ m/s}$$
$$\text{solids loading} = 6.57 \text{ kg solid/kg air}$$
$$\text{total system pressure loss} = 0.512 \text{ bar}$$

## WORKED EXAMPLE 6.2

A 20 m long standpipe carrying a Group A solids at a rate of 80 kg/s is to be aerated in order to maintain fluidized flow with a voidage in the range 0.5–0.53. Solids enter the top of the standpipe at a voidage of 0.53. The pressure and gas density at the top of the standpipe are 1.3 bar (abs) and 1.0 kg/m³ respectively. The particle density of the solids is 1200 kg/m³.
Determine the aeration positions and rates.

## Solution

From Equation (6.29), pressure ratio,

$$\frac{p_2}{p_1} = \frac{(1 - 0.50)}{0.50} \frac{0.53}{(1 - 0.53)} = 1.128$$

Therefore, $p_2 = 1.466$ bar (abs)
Pressure difference, $p_2 - p_1 = 0.166 \times 10^5$ Pa.
Hence, from Equation (6.30) (with $\varepsilon_a = [0.5 + 0.53]/2 = 0.515$),

$$\text{length to first aeration point, } H = \frac{0.166 \times 10^5}{1200 \times (1 - 0.515) \times 9.81} = 2.91 \text{ m}$$

Assuming ideal gas behaviour, density at level 2, $\rho_{f_2} = \rho_{f_1} \left( \frac{p_2}{p_1} \right) = 1.128 \text{ kg/m}^3$

Applying Equation (6.34), aeration gas mass flow at first aeration point,

$$M_{f_2} = \frac{0.53}{(1 - 0.53)} \frac{80}{1200} (1.128 - 1.0) = 0.0096 \text{ kg/s}$$

The above calculation is repeated in order to determine the position and rates of subsequent aeration points. The results are summarized below:

|                                          | First point | Second point | Third point | Fourth point | Fifth point |
|------------------------------------------|-------------|--------------|-------------|--------------|-------------|
| Distance from top of standpipe (m)       | 2.91        | 6.18         | 9.88        | 14.04        | 18.75       |
| Aeration rate (kg/s)                     | 0.0096      | 0.0108       | 0.0122      | 0.0138       | 0.0155      |
| Pressure at aeration point (bar)         | 1.47        | 1.65         | 1.86        | 2.10         | 2.37        |

## WORKED EXAMPLE 6.3

A 10 m long vertical standpipe of inside diameter 0.1 m transports solids at flux of 100 kg/m² s from an upper vessel which is held at a pressure 1.0 bar to a lower vessel held at 1.5 bar. The particle density of the solids is 2500 kg/m³ and the surface-volume mean particle size is 250 μm. Assuming that the voidage is constant along the standpipe and equal to 0.50, and that the effect of pressure change may be ignored, determine the direction and flow rate of gas passing between the vessels. (Properties of gas in the system: density, 1 kg/m³; viscosity $2 \times 10^{-5}$ Pa s.)

## Solution

First check that the solids are moving in packed bed flow. We do this by comparing the actual pressure gradient with the pressure gradient for fluidization.

Assuming that in fluidized flow the apparent weight of the solids will be supported by the gas flow, Equation (6.26) gives the pressure gradient for fluidized bed flow:

$$\frac{(-\Delta p)}{H} = (1 - 0.5) \times (2500 - 1) \times 9.81 = 12258 \text{ Pa/m}$$

$$\text{Actual pressure gradient} = \frac{(1.5 - 1.0) \times 10^5}{10} = 5000 \text{ Pa/m}$$

Since the actual pressure gradient is well below that for fluidized flow, the standpipe is operating in packed bed flow.

The pressure gradient in packed bed flow is generated by the upward flow of gas through the solids in the standpipe. The Ergun equation (Equation (6.25)) provides the relationship between gas flow and pressure gradient in a packed bed.

Knowing the required pressure gradient, the packed bed voidage and the particle and gas properties, Equation (6.25) can be solved for $|U_{rel}|$, the magnitude of the relative gas velocity:

Ignoring the negative root of the quadratic, $|U_{rel}| = 0.1026 \text{ m/s}$

We now adopt a sign convention for velocities. For standpipes it is convenient to take downward velocities as positive. In order to create the pressure gradient in the required direction, the gas must flow upwards relative to the solids. Hence, $U_{rel}$ is negative:

$$U_{rel} = -0.1026 \text{ m/s}$$

From the continuity for the solids (Equation (6.11)),

$$\text{solids flux,} = \frac{M_p}{A} = U_p(1 - \varepsilon)\rho_p$$

The solids flux is given as $100 \text{ kg/m}^2 \text{ s}$ and so

$$U_p = \frac{100}{(1 - 0.5) \times 2500} = 0.08 \text{ m/s}$$

Solids flow is downwards, so $U_p = +0.08 \text{ m/s}$

The relative velocity, $U_{rel} = U_f - U_p$

hence, actual gas velocity, $U_f = -0.1026 + 0.08 = -0.0226 \text{ m/s}$ (upwards)

Therefore the gas flows *upwards* at a velocity of $0.0226 \text{ m/s}$ relative to the standpipe walls. The superficial gas velocity is therefore:

$$U = \varepsilon U_f = -0.0113 \text{ m/s}$$

From the continuity for the gas (Equation (6.12)) mass flow rate of gas,

$$M_f = \varepsilon U_f \rho_f A$$

$$= -8.9 \times 10^{-5} \text{ kg/s}.$$

So for the standpipe to operate as required, $8.9 \times 10^{-5} \text{ kg/s}$ of gas must flow from the lower vessel to the upper vessel.

## EXERCISES

**6.1** Design a positive pressure dilute-phase pneumatic transport system to carry 500 kg/h of a powder of particle density 1800 kg/m³ and mean particle size 150 μm across a horizontal distance of 100 m and a vertical distance of 20 m using ambient air. Assume that the pipe is smooth, that four 90° bends are required and that the allowable pressure loss is 0.7 bar.

(Answer: 50 mm diameter pipe gives total pressure drop of 0.55 bar; superficial gas velocity 13.8 m/s)

**6.2** It is required to use an existing 50 mm inside diameter vertical smooth pipe as lift line to transfer 2000 kg/h of sand of mean particle size 270 μm and particle density 2500 kg/m³ to a process 50 m above the solids feed point. A blower is available which is capable of delivering 60 m³/h of ambient air at a pressure of 0.3 bar. Will the system operate as required?

(Answer: Using a superficial gas velocity of 8.49 m/s ($= 1.55 \times U_{CH}$) the total pressure drop is 0.344 bar. System will not operate as required since allowable $\Delta p = 0.3$ bar)

**6.3** Design a negative pressure dilute-phase pneumatic transport system to carry 700 kg/h of plastic spheres of particle density 1000 kg/m³ and mean particle size 1 mm between two points in a factory separated by a vertical distance of 15 m and a horizontal distance of 80 m using ambient air. Assume that the pipe is smooth, that five 90° bends are required and that the allowable pressure loss is 0.4 bar.

(Answer: Using a superficial gas velocity of 16.4 m/s in a pipe of inside diameter 40 mm, the total pressure drop is 0.38 bar.)

**6.4** A 25 m long standpipe carrying Group A solids at a rate of 75 kg/s is to be aerated in order to maintain fluidized flow with a voidage in the range 0.50–0.55. Solids enter the top of the standpipe at a voidage of 0.55. The pressure and gas density at the top of the standpipe are 1.4 bar (abs) and 1.1 kg/m³ respectively. The particle density of the solids is 1050 kg/m³.
   Determine the aeration positions and rates.

(Answer: Positions: 6.36 m, 14.13 m, 23.6 m. Rates: 0.0213 kg/s, 0.0261 kg/s, 0.0319 kg/s)

**6.5** A 15 m long standpipe carrying Group A solids at a rate of 120 kg/s is to be aerated in order to maintain fluidized flow with a voidage in the range 0.50–0.54. Solids enter the top of the standpipe at a voidage of 0.54. The pressure and gas density at the top of the standpipe are 1.2 bar (abs) and 0.9 kg/m³ respectively. The particle density of the solids is 1100 kg/m³.
   Determine the aeration positions and rates. What is the pressure at the lowest aeration point?

(Answer: Positions: 4.03 m, 8.76 m, 14.3 m. Rates: 0.0200 kg/s, 0.0235 kg/s, 0.0276 kg/s. Pressure: 1.94 bar)

**6.6** A 5 m long vertical standpipe of inside diameter 0.3 m transports solids at flux of 500 kg/m$^2$ s from an upper vessel which is held at a pressure of 1.25 bar to a lower vessel held at 1.6 bar. The particle density of the solids is 1800 kg/m$^3$ and the surface-volume mean particle size is 200 μm. Assuming that the voidage is 0.48 and is constant along the standpipe, and that the effect of pressure change may be ignored, determine the direction and flow rate of gas passing between the vessels. (Properties of gas in the system: density, 1.5 kg/m$^3$; viscosity 1.9 × 10$^{-5}$ Pa s.)

(Answer: 0.023 kg/s downwards)

**6.7** A vertical standpipe of inside diameter 0.3 m transports solids at a flux of 300 kg/m$^2$ s from an upper vessel which is held at a pressure 2.0 bar to a lower vessel held at 2.72 bar. The particle density of the solids is 2000 kg/m$^3$ and the surface-volume mean particle size is 220 μm. The density and viscosity of the gas in the system are 2.0 kg/m$^3$ and 2 × 10$^{-5}$ Pa s respectively.

   Assuming that the voidage is 0.47 and constant along the standpipe, and that the effect of pressure change may be ignored,

(a)  Determine the minimum standpipe length required to avoid fluidized flow.

(b)  Given that the actual standpipe is 8 m long, determine the direction and flow rate of gas passing between the vessels.

(Answer: (a) 6.92 m; (b) 0.0114 kg/s downwards)

# 7

# Separation of Particles from a Gas: Gas Cyclones

There are many cases during the processing and handling of particulate solids when particles are required to be separated from suspension in a gas. We saw in Chapter 5 that in fluidized bed processes the passage of gas through the bed entrains fine particles. These particles must be removed from the gas and returned to the bed before the gas can be discharged or sent to the next stage in the process. Keeping the very small particles in the fluid bed may be crucial to the successful operation of the process, as is the case in fluid catalytic cracking of oil.

In Chapter 6, we saw how a gas may be used to transport powders within a process. The efficient separation of the product from the gas at the end of the transport line plays an important part in the successful application of this method of powder transportation. In the combustion of solid fuels, fine particles of fuel ash become suspended in the combustion gases and must be removed before the gases can be discharged to the environment.

In any application, the size of the particles to be removed from the gas determine, to a large extent, the method to be used for their separation. Generally speaking, particles larger than about 100 μm can be separated easily by gravity settling. For particles less than 10 μm more energy intensive methods such as filtration, wet scrubbing and electrostatic precipitation must be used. Figure 7.1 shows typical grade efficiency curves for gas–particle separation devices. The grade efficiency curve describes how the separation efficiency of the device varies with particle size. In this chapter we will focus on the device known as the cyclone separator or cyclone. Gas cyclones are generally not suitable for separation involving suspensions with a large proportion of particles less than 10 μm. They are best suited as primary separation devices and for relatively coarse particles, with an electrostatic precipitator or fabric filter being used downstream to remove very fine particles.

Readers wishing to know more about other methods of gas–particle separa-

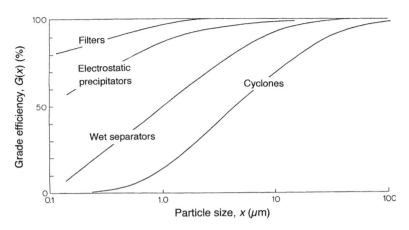

**Figure 7.1**   Typical grade efficiency curves for gas-particle separators

tion and about the choice between methods are referred to Svarovsky (1981, 1990) and Perry and Green (1984).

## 7.1   GAS CYCLONES – DESCRIPTION

The most common type of cyclone is known as the reverse flow type, shown in Figure 7.2. Inlet gas is brought tangentially into the cylindrical section and a strong vortex is thus created inside the cyclone body. Particles in the gas are subjected to centrifugal forces which move them radially outwards, against the inward flow of gas and towards the inside surface of the cyclone on which the solids separate. The direction of flow of the vortex reverses near the bottom of the cylindrical section and the gas leaves the cyclone via the outlet in the top

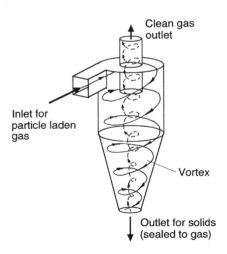

**Figure 7.2**   Schematic diagram of a reverse flow cyclone separator

(the solids outlet is sealed to gas). The solids at the wall of the cyclone are pushed downwards by the outer vortex and out of the solids exit. Gravity has been shown to have little effect on the operation of the cyclone.

## 7.2 FLOW CHARACTERISTICS

Rotational flow in the forced vortex within the cyclone body gives rise to a radial pressure gradient. This pressure gradient, combined with the frictional pressure losses at the gas inlet and outlet and losses due to changes in flow direction, make up the total static pressure drop. This static pressure drop, measured between the inlet and the gas outlet, is usually proportional to the square of gas flow rate through the cyclone. A resistance coefficient, the Euler number $Eu$, relates the cyclone pressure drop $\Delta p$ to a characteristic velocity $v$:

$$Eu = \Delta p/(\rho_f v^2/2) \tag{7.1}$$

where $\rho_f$ is the gas density.

The characteristic velocity $v$ can be defined for gas cyclones in various ways but the simplest and most appropriate definition is based on the cross-section of the cylindrical body of the cyclone, so that:

$$v = 4q/(\pi D^2) \tag{7.2}$$

where $q$ is the gas flow rate and $D$ is the cyclone inside diameter.

The Euler number represents the ratio of pressure forces to the inertial forces acting on a fluid element. Its value is practically constant for a given cyclone geometry, independent of the cyclone body diameter (see Section 7.4).

## 7.3 EFFICIENCY OF SEPARATION

### 7.3.1 Total Efficiency and Grade Efficiency

Consider a cyclone to which the solids mass flow rate is $M$, the mass flow discharged from the solids exit orifice is $M_c$ (known as the coarse product) and the solids mass flow rate leaving with the gas is $M_f$ (known as the fine product). The total material balance on the solids over this cyclone may be written:

$$\text{Total:} \quad M = M_f + M_c \tag{7.3}$$

and the 'component' material balance for each particle size $x$ (assuming no breakage or growth or particles within the cyclone) is

$$\text{Component:} \quad M(dF/dx) = M_f(dF_f/dx) + M_c(dF_c/dx) \tag{7.4}$$

where, $dF/dx$, $dF_f/dx$ and $dF_c/dx$ are the differential frequency size distributions by mass (i.e. mass fraction of size $x$) for the feed, fine product and coarse product respectively. $F$, $F_f$ and $F_c$ are the cumulative frequency size distributions by mass (mass fraction less than size $x$) for the feed, fine product and coarse product respectively. Refer to Chapter 3 for further details on representations of particle size distributions.

The total efficiency of separation of particles from gas, $E_T$, is defined as the fraction of the total feed which appears in the coarse product collected, i.e.

$$E_T = M_c/M \tag{7.5}$$

The efficiency with which the cyclone collects particles of a certain size is described by the grade efficiency, $G(x)$, which is defined as

$$G(x) = \frac{\text{mass of solids of size } x \text{ in coarse product}}{\text{mass of solids of size } x \text{ in feed}} \tag{7.6}$$

Using the notation for size distribution described above,

$$G(x) = \frac{M_c(dF_c/dx)}{M(dF/dx)} \tag{7.7}$$

Combining with Equation (7.5), we find an expression linking grade efficiency with total efficiency of separation:

$$G(x) = E_T \frac{(dF_c/dx)}{(dF/dx)} \tag{7.8}$$

From Equations (7.3)–(7.5), we have

$$(dF/dx) = E_T(dF_c/dx) + (1 - E_T)(dF_f/dx) \tag{7.9}$$

Equation (7.9) relates the size distributions of the feed (no subscript), the coarse product (subscript c) and the fine product (subscript f). In cumulative form this becomes

$$F = E_T F_c + (1 - E_T)F_f \tag{7.10}$$

### 7.3.2 Simple Theoretical Analysis for the Gas Cyclone Separator

Referring to Figure 7.3, consider a reverse flow cyclone with a cylindrical section of radius $R$. Particles entering the cyclone with the gas stream are forced into circular motion. The net flow of gas is radially inwards towards the central gas outlet. The forces acting on a particle following a circular path are drag, buoyancy and centrifugal force. The balance between these forces determines the equilibrium orbit adopted by the particle. The drag force is

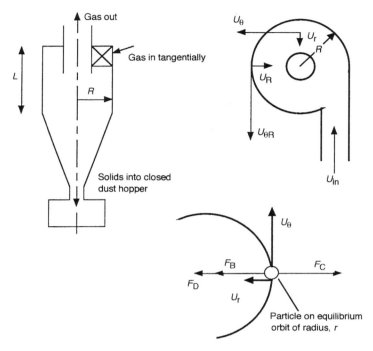

**Figure 7.3**  Reverse flow cyclone – a simple theory for separation efficiency

caused by the inward flow of gas past the particle and acts radially inwards. Consider a particle of diameter $x$ and density $\rho_p$ following an orbit of radius $r$ in a gas of density, $\rho_f$ and viscosity $\mu$. Let the tangential velocity of the particle be $U_\theta$ and the radial inward velocity of the gas be $U_r$. If we assume that Stokes' law applies under these conditions then the drag force is given by

$$F_D = 3\pi x \mu U_r \tag{7.11}$$

The centrifugal and buoyancy forces acting on the particle moving with a tangential velocity component $U_\theta$ at radius $r$ are respectively

$$F_C = \frac{\pi x^3}{6} \rho_P \frac{U_\theta^2}{r} \tag{7.12}$$

$$F_B = \frac{\pi x^3}{6} \rho_f \frac{U_\theta^2}{r} \tag{7.13}$$

Under the action of these forces the particle moves inwards or outwards until the forces are balanced and the particle assumes its equilibrium orbit. At this point,

$$F_C = F_D + F_B \tag{7.14}$$

and so

$$x^2 = \frac{18\mu}{(\rho_p - \rho_f)} \left(\frac{r}{U_\theta^2}\right) U_r \qquad (7.15)$$

To go any further we need a relationship between $U_\theta$ and the radius $r$ for the vortex in a cyclone. Now for a rotating solid body, $U_\theta = r\omega$, where $\omega$ is the angular velocity and for a free vortex $U_\theta r = $ constant. For the confined vortex inside the cyclone body it is has been found experimentally that the following holds approximately:

$$U_\theta r^{1/2} = \text{constant}$$

hence,

$$U_\theta r^{1/2} = U_{\theta R} R^{1/2} \qquad (7.16)$$

If we also assume uniform flow of gas towards the central outlet, then we are able to derive the radial variation in the radial component of gas velocity, $U_r$:

$$\text{gas flow rate,} \quad q = 2\pi r L U_r = 2\pi R L U_R \qquad (7.17)$$

hence,

$$U_R = U_r(r/R) \qquad (7.18)$$

Combining equations (7.16) and (7.18) with equation (7.15), we find

$$x^2 = \frac{18\mu}{(\rho_p - \rho_f)} \frac{U_R}{U_{\theta R}^2} r \qquad (7.19)$$

where $r$ is the radius of the equilibrium orbit for a particle of diameter $x$.

If we assume that all particles with an equilibrium orbit radius greater than or equal to the cyclone body radius will be collected, then substituting $r = R$ in Equation (7.19) we derive the expression below for the critical particle diameter for separation, $x_{\text{crit}}$:

$$x_{\text{crit}}^2 = \frac{18\mu}{(\rho_p - \rho_f)} \frac{U_R}{U_{\theta R}^2} R \qquad (7.20)$$

The values of the radial and tangential velocity components at the cyclone wall, $U_R$ and $U_{\theta R}$, in Equation (7.20) may be found from a knowledge of the cyclone geometry and the gas flow rate.

This analysis predicts an ideal grade efficiency curve shown in Figure 7.4. All particles of diameter $x_{\text{crit}}$ and greater are collected and all particles of size less than $x_{\text{crit}}$ are not collected.

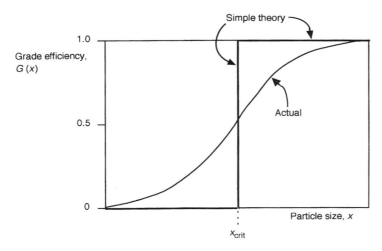

**Figure 7.4** Theoretical and actual grade efficiency curves

### 7.3.3 Cyclone Grade Efficiency in Practice

In practice, gas velocity fluctuations and particle–particle interactions result in some particles larger than $x_{crit}$ being lost and some particles smaller than $x_{crit}$ being collected. Consequently, in practice the cyclone does not achieve such a sharp cut-off as predicted by the theoretical analysis above. In common with other separation devices in which body forces are opposed by drag forces, the grade efficiency curve for gas cyclones is usually S-shaped.

For such a curve, the particle size for which the grade efficiency is 50%, $x_{50}$ is often used as a single number measurement of the efficiency of the cyclone. $x_{50}$ is also know as the equiprobable size since it is that size of particle which has a 50% probability of appearing in the coarse product. This also means that, in a large population of particles, 50% of the particles of this size will appear in the coarse product. $x_{50}$ is sometimes simply referred to as the cut size of the cyclone (or other separation device).

The concept of $x_{50}$ cut size is useful where the efficiency of a cyclone is to be expressed as a single number independent of the feed solid size distribution, such as in scale-up calculation.

### 7.4 SCALE-UP OF CYCLONES

The scale-up of cyclones is based on a dimensionless group, the Stokes number, which characterizes the separation performance of a family of geometrically similar cyclones. The Stokes number $Stk_{50}$ is defined as

$$Stk_{50} = \frac{x_{50}^2 \rho_P v}{18\mu D} \qquad (7.21)$$

where $\mu$ is gas viscosity, $\rho_p$ is solids density, $v$ is the characteristic velocity defined by Equation (7.2) and $D$ is the diameter of the cyclone body. The physical significance of the Stokes number is that it is a ratio of the centrifugal force (less buoyancy) to the drag force, both acting on a particle of size $x_{50}$. Readers will note the similarity between our theoretical expression of Equation (7.20) and the Stokes number of Equation (7.21). There is therefore some theoretical justification for the use of the Stokes number in scale-up.

For large industrial cyclones the Stokes number, like the Euler number defined previously, is independent of Reynolds number. For suspensions of concentration less than about $5 \, g/m^3$, the Stokes and Euler numbers are usually constant for a given cyclone geometry (i.e. a set of geometric proportions relative to cyclone diameter $D$). The geometries and values of $Eu$ and $Stk_{50}$ for two common industrial cyclones, the Stairmand high efficiency (HE) and the Stairmand high rate (HR) are given in Figure 7.5.

The use of the two dimensionless groups $Eu$ and $Stk_{50}$ in cyclone scale-up and design is demonstrated in the worked examples at the end of this chapter.

As can be seen from Equation (7.21), the separation efficiency is described only by the cut size $x_{50}$ and no regard is given to the shape of the grade efficiency curve. If the whole grade efficiency curve is required in performance calculations, it may be generated around the given cut size using plots or analytical functions of a generalized grade efficiency function available from the literature or from previously measured data. For example, Perry (1984) gives the grade efficiency expression:

$$\text{grade efficiency} = \frac{(x/x_{50})^2}{[1 + (x/x_{50})^2]} \tag{7.22}$$

for a reverse flow cyclone with the geometry:

| A | B | C | E | J | K | N |
|-----|-----|-----|------|-------|-----|-----|
| 4.0 | 2.0 | 2.0 | 0.25 | 0.625 | 0.5 | 0.5 |

(Letters refer to the cyclone geometry diagram shown in Figure 7.5)

This expression gives rise to the grade efficiency curve shown in Figure 7.6 for an $x_{50}$ cut size of 5 µm. Very little is known how the shape of the grade efficiency curve is affected by operating pressure drop, cyclone size or design, and feed solids concentration.

## 7.5   RANGE OF OPERATION

One of the most important characteristics of gas cyclones is the way in which their efficiency is affected by pressure drop (or flow rate). For a particular cyclone and inlet particle concentration, total efficiency of separation and

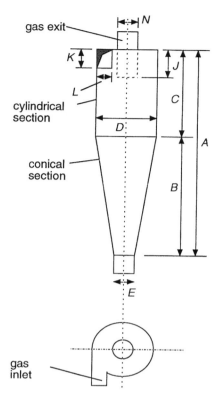

gas exit

cylindrical
section

conical
section

gas
inlet

HE – high efficiency Stairmand cyclone
$Stk_{50} = 1.4 \times 10^{-4}$
$Eu = 320$

HR – high flowrate Stairmand cyclone
$Stk_{50} = 6 \times 10^{-3}$
$Eu = 46$

Dimension relative to diameter D

| Cyclone type | A | B | C | E | J | L | K | N |
|---|---|---|---|---|---|---|---|---|
| Stairmand, H.E. | 4.0 | 2.5 | 1.5 | 0.375 | 0.5 | 0.2 | 0.5 | 0.5 |
| Stairmand, H.R. | 4.0 | 2.5 | 1.5 | 0.575 | 0.875 | 0.375 | 0.75 | 0.75 |

**Figure 7.5** Geometries and Euler and Stokes numbers for two common cyclones

pressure drop vary with gas flow rate as shown in Figure 7.7. Theory predicts that efficiency increases with increasing gas flowrate. However, in practice, the total efficiency curve falls away at high flowrates because re-entrainment of separated solids increases with increased turbulence at high velocities. Optimum operation is achieved somewhere between points A and B, where maxi-

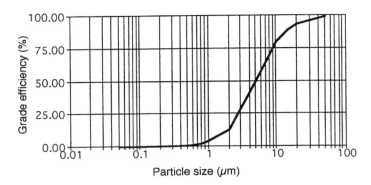

**Figure 7.6** Grade efficiency curve described by Equation (7.22) for a cut size $x_{50} = 5\ \mu m$

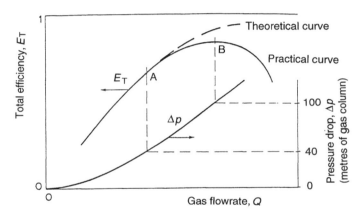

**Figure 7.7** Total separation efficiency and pressure drop versus gas flowrate through a reverse flow cyclone

mum total separation efficiency is achieved with reasonable pressure loss (and hence power consumption). The position of point B changes only slightly for different dusts. Correctly designed and operated cyclones should operate at pressure drops within a recommended range; and this, for most cyclone designs operated at ambient conditions, is between 50 and 150 mm of water gauge (WG) (approximately from 500 to 1500 Pa). Within this range, the total separation efficiency $E_T$ increases with applied static pressure drop, in accordance with the inertial separation theory shown above.

Above the top limit the total efficiency no longer increases with increasing pressure drop and it may actually decline due to re-entrainment of dust from the dust outlet orifice. It is, therefore, wasteful of energy to operate cyclones above the limit. At pressure drops below the bottom limit, the cyclone represents little more than a settling chamber, giving low efficiency due to low velocities within it which may not be capable of generating a stable vortex.

## 7.6   SOME PRACTICAL DESIGN AND OPERATION DETAILS

The following practical considerations for design and operation of reverse flow gas cyclones are among those listed by Svarovsky (1986).

### 7.6.1   Effect of Dust Loading on Efficiency

One of the important operating variables affecting total efficiency is the concentration of particles in the suspension (known as the dust loading). Generally, high dust loadings (above about 5 g/m$^3$) lead to higher total separation efficiencies due to particle enlargement through agglomeration of particles (caused, for example, by the effect of humidity).

### 7.6.2   Cyclone Types

The many reverse flow cyclone designs available today may be divided into two main groups: *high efficiency* designs (e.g. Stairmand HE) and the *high rate* designs (e.g. Stairmand HR). High efficiency cyclones give high recoveries and are characterized by relatively small inlet and gas outlet orifices. The high rate designs have lower total efficiencies, but offer low resistance to flow so that a unit of a given size will give much higher gas capacity than a high efficiency design of the same body diameter. The high rate cyclones have large inlets and gas outlets, and are usually shorter. The geometries and values of $Eu$ and $Stk_{50}$ for two common cyclones, the Stairmand high efficiency (HE) and the Stairmand high rate (HR) are given in Figure 7.5.

For well-designed cyclones there is a direct correlation between $Eu$ and $Stk_{50}$. High values of the resistance coefficient usually lead to low values of $Stk_{50}$ (therefore low cut sizes and high efficiencies), and vice versa. The general trend can be described by the following *approximate* empirical correlation:

$$Eu = \sqrt{\frac{12}{Stk_{50}}} \qquad (7.23)$$

### 7.6.3   Abrasion

Abrasion in gas cyclones is an important aspect of cyclone performance and it is affected by the way cyclones are installed and operated as much as by the material construction and design. Materials of construction are usually steels of different grades, sometimes lined with rubber, refractory lining or other material. Within the cyclone body there are two critical zones for abrasion: in the cylindrical part just beyond the inlet opening and in the conical part near the dust discharge.

### 7.6.4   Attrition of Solids

Attrition or break-up of solids is known to take place on collection in gas cyclones but little is known about how it is related to particle properties, although large particles are more likely to be affected by attrition than finer fractions. Attrition is most detectable in recirculating systems such as fluidized beds where cyclones are used to return the carry-over material back to the bed (see Chapter 5). The complete inventory of the bed may pass through the cyclones many times per hour and the effect of attrition is thus exaggerated.

### 7.6.5   Blockages

Blockages, usually caused by overloading of the solids outlet orifice, is one of the most common causes of failure in cyclone operation. The cyclone cone rapidly fills up with dust, the pressure drop increases and efficiency falls dramatically. Blockages arise due to mechanical defects in the cyclone body (bumps on the cyclone cone, protruding welds or gasket) or changes in chemical or physical properties of the solids (e.g. condensation of water vapour from the gas onto the surface of particles).

### 7.6.6   Discharge Hoppers and Diplegs

The design of the solids discharge is important for correct functioning of a gas cyclone. If the cyclone operates under vacuum, any inward leakages of air at the discharge end cause particles to be re-entrained and this leads to a sharp decrease in separation efficiency. If the cyclone is under pressure, outward leakages may cause a slight increase in separation efficiency, but also results in loss of product and pollution of the local environment. It is therefore best to keep the solids discharge as gas-tight as possible.

The strong vortex inside a cyclone reaches into the space underneath the solids outlet and it is important that no powder surface is allowed to build up to within at least one cyclone diameter below the underflow orifice. A conical vortex breaker positioned just under the dust discharge orifice may be used to prevent the vortex from intruding into the discharge hopper below. Some cyclone manufacturers use a 'stepped' cone to counter the effects of re-entrainment and abrasion, and Svarovsky (1984) demonstrated the value of this design feature.

In fluidized beds with internal cyclones, 'diplegs' are used to return the collected entrained particles into the fluidized bed. Diplegs are vertical pipes connected directly to the solids discharge orifice of the cyclone extending down to below the fluidized bed surface. Particles discharged from the cyclone collect as a moving settled suspension in the lower part of the dipleg before it enters the bed. The level of the settled suspension in the dipleg is always higher than the fluidized bed surface and it provides a necessary resistance to minimize both the flow of gas up the dipleg and the consequent reduction of cyclone efficiency.

## 7.6.7 Cyclones in Series

Connecting cyclones in series is often done in practice to increase recovery. Usually the primary cyclone would be of medium or low efficiency design and the secondary and subsequent cyclones of progressively more efficient design or smaller diameter.

## 7.6.8 Cyclones in Parallel

The $x_{50}$ cut size achievable for a given cyclone geometry and operating pressure drop decreases with decreasing cyclone size (see Equation (7.21)). The size a single cyclone for treating a given volume flowrate of gas is determined by that gas flowrate (Equations (7.1) and (7.2)). For large gas flowrates the resulting cyclone may be so large that the $x_{50}$ cut size is unacceptably high. The solution is to split the gas flow into several smaller cyclones operating in parallel. In this way, both the operating pressure drop and $x_{50}$ cut size requirements can be achieved. The worked examples at the end of the chapter demonstrate how the number and diameter of cyclones in parallel are estimated.

## 7.7  WORKED EXAMPLES

### WORKED EXAMPLE 7.1 – DESIGN OF A CYCLONE

Determine the diameter and number of gas cyclones required to treat 2 m³/s of ambient air (viscosity $= 18.25 \times 10^{-6}$ Pa s, density $= 1.2$ kg/m³) laden with solids of density 1000 kg/m³ at a suitable pressure drop and with a cut size of 4 μm. Use a Stairmand HE (high efficiency) cyclone for which $Eu = 320$ and $Stk_{50} = 1.4 \times 10^{-4}$.

$$\text{Optimum pressure drop} = 100 \, \text{m gas}$$

$$= 100 \times 1.2 \times 9.81 \, \text{Pa}$$

$$= 1177 \, \text{Pa}$$

*Solution*

From Equation (7.1),

$$\text{characteristic velocity, } v = 2.476 \, \text{m/s}$$

Hence, from Equation (7.2), diameter of cyclone, $D = 1.014$ m
With this cyclone, using Equation (7.21), cut size, $x_{50} = 4.34$ μm
This is too high and we must therefore opt for passing the gas through several smaller cyclones in parallel.

Assuming that $n$ cyclones in parallel are required and that the total flow is evenly split, then for each cyclone the flow rate will be $q = 2/n$.
Therefore from Equations (7.1) and (7.2), new cyclone diameter, $D = 1.014/n^{0.5}$
Substituting in Equation (7.21) for $D$, the required cut size and $v$ (2.476 m/s, as originally calculated, since this is determined solely by the pressure drop requirement), we find that

$$n = 1.386.$$

We will therefore need two cyclones. Now with $n = 2$, we recalculate the cyclone diameter from $D = 1.014/n^{0.5}$ and the actual achieved cut size from Equation (7.21).
Thus, $D = 0.717$ m, and using this value for $D$ in Equation (7.21) together with required cut size and $v = 2.476$ m/s, we find that the actual cut size is 3.65 μm.
Therefore, two 0.717 m diameter Stairmand HE cyclones in parallel will give cut size of 3.65 μm using a pressure drop of 1177 Pa.

## WORKED EXAMPLE 7.2

Tests on a reverse flow gas cyclone give the results shown in the table below.

| Size range (μm) | 0–5 | 5–10 | 10–15 | 15–20 | 20–25 | 25–30 |
|---|---|---|---|---|---|---|
| Feed size analysis, $m$ (g) | 10 | 15 | 25 | 30 | 15 | 5 |
| Course product size analysis, $m_c$ (g) | 0.1 | 3.53 | 18.0 | 27.3 | 14.63 | 5.0 |

(a)  From these results determine the total efficiency of the cyclone.

(b)  Plot the grade efficiency curve and hence show that the $x_{50}$ cut size is 10 μm.

(c)  The dimensionless constants describing this cyclone are: $Eu = 384$ and $Stk_{50} = 1 \times 10^{-3}$. Determine the diameter and number of cyclones to be operated in parallel to achieve this cut size when handling 10 m³/s of a gas of density 1.2 kg/m³ and viscosity $18.4 \times 10^{-6}$ Pa s, laden with dust of particle density 2500 kg/m³. The available pressure drop is 1200 Pa.

(d)  What is the actual cut size of your design?

## Solution

(a) From the test results:
Mass of feed, $M = 10 + 15 + 25 + 30 + 15 + 5 = 100$ g.
Mass of coarse product, $M_c = 0.1 + 3.53 + 18.0 + 27.3 + 14.63 + 5.0 = 68.56$ g.

Therefore, from Equation (7.5), total efficiency,

$$E_T = \frac{M_c}{M} = 0.6586 \text{ (or } 68.56\%)$$

(b) From Equation (7.7), grade efficiency,

$$G(x) = \frac{M_c}{M}\frac{dF_c/dx}{dF/dx} = E_T\frac{dF_c/dx}{dF/dx}$$

In this case, $G(x)$ may be obtained directly from the results table as

$$G(x) = \frac{m_c}{m}$$

And so, the grade efficiency curve data becomes.

| Size range (µm) | 0–5 | 5–10 | 10–15 | 15–20 | 20–25 | 25–30 |
|---|---|---|---|---|---|---|
| $G(x)$ | 0.01 | 0.235 | 0.721 | 0.909 | 0.975 | 1.00 |

Plotting these data gives $x_{50} = 10$ µm, as may be seen from Figure 7W.1.1. For interest, we can calculate the size distributions of the feed, $dF/dx$ and the coarse product, $dF_c/dx$:

| Size range, x (µm) | 0–5 | 5–10 | 10–15 | 15–20 | 20–25 | 25–30 |
|---|---|---|---|---|---|---|
| $dF_c/dx$ | 0.001 46 | 0.0515 | 0.263 | 0.398 | 0.2134 | 0.0729 |
| $dF/dx$ | 0.1 | 0.15 | 0.25 | 0.30 | 0.15 | 0.05 |

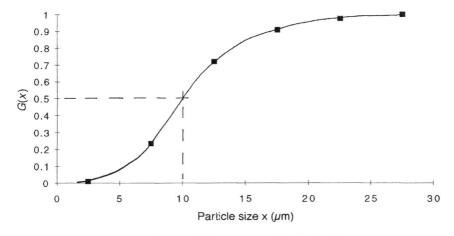

**Figure 7.W1.1**   Grade efficiency curve for Worked Example 7.2

We can then verify the calculated $G(x)$ values. For example, in the size range, 10–15,

$$G(x) = E_T \frac{\mathrm{d}F_c/\mathrm{d}x}{\mathrm{d}F/\mathrm{d}x} = 0.6856 \times \frac{0.263}{0.25} = 0.721$$

(c) Using Equation (7.1), noting that the allowable pressure is 1200 Pa, we calculate the characteristic velocity, $v$:

$$v = \sqrt{\frac{2\Delta p}{Eu\rho_f}} = \sqrt{\frac{2 \times 1200}{384 \times 1.2}} = 2.282 \text{ m/s}$$

If we have $n$ cyclones in parallel then assuming even distribution of the gas between the cyclones, flowrate to each cyclone, $q = Q/n$ and from Equation (7.2),

$$D = \sqrt{\frac{4Q}{n\pi v}} = \sqrt{\frac{4 \times 10}{n\pi \times 2.282}} = \frac{2.362}{\sqrt{n}}$$

Now substitute this expression for $D$ and the required cut size $x_{50}$ in Equation (7.21) for $Stk_{50}$:

$$Stk_{50} = \frac{x_{50}^2 \rho_p v}{18\mu D}$$

$$1 \times 10^{-3} = \frac{(10 \times 10^{-6})^2 \times 2500 \times 2.282}{18 \times 18.4 \times 10^{-6} \times (2.362/\sqrt{n})}$$

giving $n = 1.88$.

We therefore require two cyclones. With two cyclones, using all of the allowable pressure drop the characteristic velocity will be the same (2.282 m/s) and the required cyclone diameter may be calculated from the expression derived above:

$$D = \frac{2.362}{\sqrt{n}}, \text{ giving } D = 1.67 \text{ m}.$$

(d) The actual cut size achieved with two cyclones is calculated from Equation (7.21) with $D = 1.67$ m and $v = 2.282$ m/s:

$$\text{actual cut size, } x_{50} = \sqrt{\frac{1 \times 10^{-3} \times 18 \times 18.4 \times 10^{-6} \times 1.67}{2500 \times 2.282}} = 9.85 \times 10^{-6} \text{ m}$$

*Summary.* Two cyclones (described by $Eu = 384$ and $Stk_{50} = 1 \times 10^{-3}$) of diameter 1.67 m and operating at a pressure drop of 1200 Pa, will achieve a equiprobable cut size of 9.85 µm.

## EXERCISES

**7.1** A gas–particle separation device is tested and gives the results shown in the table below:

| Size range (µm) | 0–10 | 10–20 | 20–30 | 30–40 | 40–50 |
|---|---|---|---|---|---|
| Range mean (µm) | 5 | 15 | 25 | 35 | 45 |
| Feed mass (kg) | 45 | 69 | 120 | 45 | 21 |
| Coarse product mass (kg) | 1.35 | 19.32 | 99.0 | 44.33 | 21.0 |

(a) Find the total efficiency of the device.

(b) Produce a plot of the grade efficiency for this device and determine the equiprobable cut size.

(Answer: (a) 61.7%; (b) 19.4 µm)

**7.2** A gas–particle separation device is tested and gives the results shown in the table below:

| Size range (µm) | 6.6–9.4 | 9.4–13.3 | 13.3–18.7 | 18.7–27.0 | 27.0–37.0 | 37.0–53.0 |
|---|---|---|---|---|---|---|
| Feed size distribution | 0.05 | 0.2 | 0.35 | 0.25 | 0.1 | 0.05 |
| Coarse product size distribution | 0.016 | 0.139 | 0.366 | 0.30 | 0.12 | 0.06 |

Given that the total mass of feed is 200 kg and the total mass of coarse product collected is 166.5 kg,

(a) Find the total efficiency of the device

(b) Determine the size distribution of the fine product.

(c) Plot the grade efficiency curve for this device and determine the equiprobable size.

(d) If this same device were fed with a material with the size distribution below, what would be the resulting coarse product size distribution?

| Size range (µm) | 6.6–9.4 | 9.4–13.3 | 13.3–18.7 | 18.7–27.0 | 27.0–37.0 | 37.0–53.0 |
|---|---|---|---|---|---|---|
| Feed size distribution | 0.08 | 0.13 | 0.27 | 0.36 | 0.14 | 0.02 |

(Answer: (a) 83.25%; (b) 0.219, 0.503, 0.271, 0.0015, 0.0006, 0.0003; (c) 10.5 µm; (d) 0.025, 0.089, 0.276, 0.422, 0.165, 0.024)

**7.3**

(a) Explain what a 'grade efficiency curve' is with reference to a gas–solids separation device and sketch an example of such a curve for a gas cyclone separator.

(b) Determine the diameter and number of Stairmand HR gas cyclones to be operated in parallel to treat 3 m$^3$/s of gas of density 0.5 kg/m$^3$ and viscosity $2 \times 10^{-5}$ Pa s carrying a dust of density 2000 kg/m$^3$. A $x_{50}$ cut size of at most 7 μm is to be achieved at a pressure drop of 1200 Pa.

(For a Stairmand HR cyclone: $Eu = 46$ and $Stk_{50} = 6 \times 10^{-3}$.)

(c) Give the actual cut size achieved by your design.

(d) A change in process conditions requirements necessitates a 50% drop in gas flowrate. What effect will this have on the cut size achieved by your design?

(Answer: (a) Two cyclones 0.43 m in diameter; (b) $x_{50} = 6.8$ μm; (c) new $x_{50} = 9.6$ μm)

**7.4**

(a) Determine the diameter and number of Stairmand HE gas cyclones to be operated in parallel to treat 1 m$^3$/s of gas of density 1.2 kg/m$^3$ and viscosity $18.5 \times 10^{-6}$ Pa s carrying a dust of density 1000 kg/m$^3$. An $x_{50}$ cut size of at most 5 μm is to be achieved at a pressure drop of 1200 Pa.
(For a Stairmand HE cyclone: $Eu = 320$ and $Stk_{50} = 1.4 \times 10^{-4}$.)

(b) Give the actual cut size achieved by your design.

(Answer: (a) One cyclone, 0.714 m in diameter; (b) $x_{50} = 3.6$ μm)

**7.5** Stairmand HR cyclones are to be used to clean up 2.5 m$^3$/s of ambient air (density 1.2 kg/m$^3$ and viscosity $18.5 \times 10^{-6}$ Pa s) laden with dust of particle density 2600 kg/m$^3$. The available pressure drop is 1200 Pa and the required cut size is to be not more than 6 μm.

(a) What size of cyclones are required?

(b) How many cyclones are needed and in what arrangement?

(c) What is the actual cut size achieved?

(Answer: (a) Diameter $= 0.311$ m; (b) five cyclones in parallel; (c) actual cut size $= 6$ μm)

# 8

# Storage and Flow of Powders – Hopper Design

## 8.1 INTRODUCTION

The short-term storage of raw materials, intermediates and products in the form of particulate solids in process plants presents problems which are often underestimated and which, as was pointed out in the introduction of this text, may frequently be responsible for production stoppages.

One common problem in such plants is the interruption of flow from the discharge orifice in the hopper, or converging section beneath a storage vessel for powders. However, a technology is available which will allow us to design such storage vessels to ensure flow of the powders when desired. Within the bounds of a single chapter it is not possible to cover all aspects of the gravity flow of unaerated powders, and so here we will confine ourselves to a study of the design philosophy to ensure flow from conical hoppers when required. The approach used is that first proposed by Jenike (1964).

## 8.2 MASS FLOW AND CORE FLOW

*Mass flow.* In perfect mass flow, all the powder in a silo is in motion whenever any of it is drawn from the outlet as shown in Figure 8.1(b). The flowing channel coincides with the walls of the silo. Mass flow hoppers are smooth and steep. Figures 8.2 (a–d) are sketches taken from a sequence of photographs of a hopper operating in mass flow. The use of alternate layers of coloured powder in this sequence clearly shows the key features of the flow pattern. Note how the powder surface remains level until it reaches the sloping section.

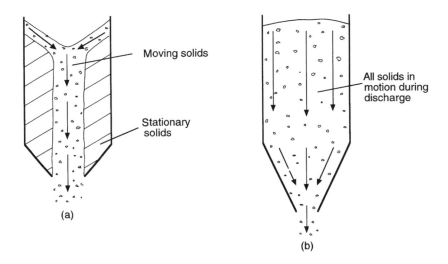

**Figure 8.1**    Mass flow and core flow in hoppers: (a) core flow; (b) mass flow

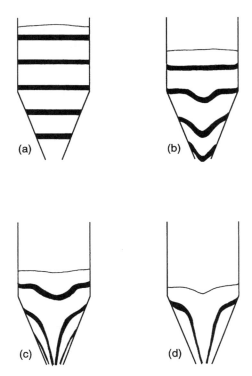

**Figure 8.2**    Sequence of sketches taken from photographs showing a mass flow pattern as a hopper empties. (The black bands are layers of coloured tracer particles)

*Core flow.* This occurs when the powder flows towards the outlet of a silo in a channel formed within the powder itself Figure 8.1(a). We will not concern ourselves with core flow silo design. Figures 8.3 (a–d) are sketches taken from a sequence of photographs of a hopper operating in core flow. Note the regions of powder lower down in the hopper are stagnant until the hopper is almost empty. The inclined surface of the powder gives rise to size segregation (see Chapter 9).

Mass flow has many advantages over core flow. In mass flow, the motion of the powder is uniform and steady state can be closely approximated. The bulk density of the discharged powder is constant and practically independent of the height in the silo. In mass flow stresses are generally low throughout the mass of solids, giving low compaction of the powder. There are no stagnant regions in the mass flow hopper. Thus the risk of product degradation is small compared with the case of the core flow hopper. The first-in–first-out flow pattern of the mass flow hopper ensures a narrow range of residence times for solids in the silo. Also, segregation of particles according to size is far less of a problem in mass flow than in core flow. Mass flow has one disadvantage which may be overriding in certain cases. Friction between the moving solids and the silo and hopper walls result in erosion of the wall, which gives rise to contamination of the solids by the

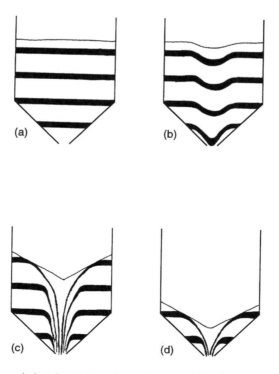

**Figure 8.3**  Sequence of sketches taken from photographs showing a core flow pattern as a hopper empties. (The black bands are layers of coloured tracer particles)

material of the hopper wall. If either contamination of the solids or serious erosion of the wall material are unacceptable, then a core flow hopper should be considered.

For conical hoppers the slope angle required to ensure mass flow depends on the powder/powder friction and the powder/wall friction. Later we will see how these are quantified and how it is possible to determine the conditions which give rise to mass flow. Note that there is no such thing as a mass flow hopper; a hopper which gives mass flow with one powder may give core flow with another.

## 8.3   THE DESIGN PHILOSOPHY

We will consider the blockage or obstruction to flow called arching and assume that if this does not occur then flow will take place, Figure 8.4. Now, in general, powders develop strength under the action of compacting stresses. The greater the compacting stress, the greater the strength developed (Figure 8.5). (Free-flowing solids such as dry coarse sand do not develop strength as the result of compacting stresses and will always flow).

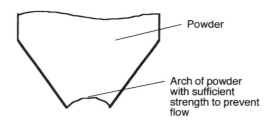

**Figure 8.4**   Arching in the flow of powder from a hopper

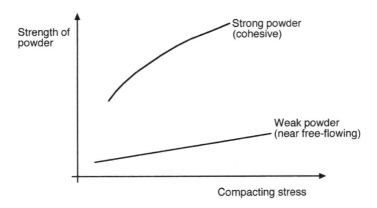

**Figure 8.5**   Variation of strength of powder with compacting stress for cohesive and free-flowing powders

## 8.3.1 Flow–No Flow Criterion

Gravity flow of a solid in a channel will take place provided the strength developed by the solids under the action of consolidating pressures is insufficient to support an obstruction to flow. An arch occurs when the strength developed by the solids is greater than the stresses acting within the surface of the arch.

## 8.3.2 The Hopper Flow Factor *ff*

The hopper flow factor, *ff*, relates the stress developed in a particulate solid with the compacting stress acting in a particular hopper. The hopper flow factor is defined in Equation (8.1):

$$ff = \frac{\sigma_C}{\sigma_D} = \frac{\text{compacting stress in the hopper}}{\text{stress developed in the powder}} \tag{8.1}$$

A high value of *ff* means low flowability since high $\sigma_C$ means greater compaction, and a low value of $\sigma_D$ means more chance of an arch forming.

The hopper flow factor *ff* depends on

- the nature of the solid;
- the nature of the wall material;
- the slope of the hopper wall.

These relationships will be quantified later.

## 8.3.3 Unconfined Yield Stress, $\sigma_y$

We are interested in the strength developed by the powder in the arch surface. Suppose that the yield stress (i.e. the stress which causes flow) of the powder in the exposed surface of the arch is $\sigma_y$. The stress $\sigma_y$ is known as the unconfined yield stress of the powder. Then if the stresses developed in the powder forming the arch are greater than the unconfined yield stress of the powder in the arch, flow will occur. That is,

$$\text{For flow} \quad \sigma_D > \sigma_y \tag{8.2}$$

Incorporating Equation (8.1), this criterion may be rewritten as:

$$\text{For flow:} \quad \frac{\sigma_C}{ff} > \sigma_y \tag{8.3}$$

### 8.3.4   Powder Flow Function

Obviously, the unconfined yield stress $\sigma_y$ of the solids varies with compacting stress ($\sigma_C$), i.e.

$$\sigma_y = fn(\sigma_C)$$

This relationship is determined experimentally and is usually presented graphically (Figure 8.6). This relationship has several different names, some of which are misleading. Here we will call it the *powder flow function*. Note that it is a function *only* of the powder properties.

### 8.3.5   Critical Conditions for Flow

From Equation (8.3), the limiting condition for flow is

$$\frac{\sigma_C}{ff} = \sigma_y$$

This may be plotted on the same axes as the powder flow function (unconfined yield stress, $\sigma_y$ and compacting stress, $\sigma_C$) in order to reveal the conditions under which flow will occur for this powder in the hopper. The limiting condition gives a straight line of slope $1/ff$. Figure 8.7 shows such a plot.

Where the powder has a yield stress greater than $\sigma_C/ff$, no flow occurs (powder flow function (a)). Where the powder has a yield stress less than $\sigma_C/ff$, flow occurs (powder flow function (c)). For powder flow function (b) there is a critical condition, where unconfined yield stress, $\sigma_y$ is equal to stress

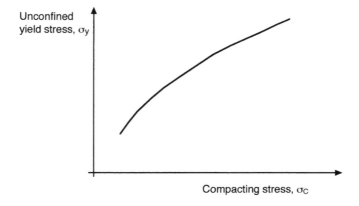

**Figure 8.6**   Powder flow function (a property of the solids only)

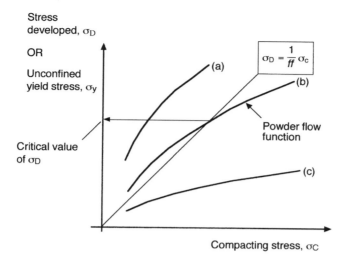

**Figure 8.7**   Determination of critical conditions for flow

developed in the powder, $\sigma_C/ff$. This gives rise to a critical value of stress, $\sigma_{crit}$, which is the critical stress developed in the surface of the arch.

If actual stress developed $< \sigma_{crit} \Rightarrow$ NO FLOW

If actual stress developed $> \sigma_{crit} \Rightarrow$ FLOW

## 8.3.6   Critical Outlet Dimension

Intuitively, for a given hopper geometry, one would expect the stress developed in the arch to increase with the span of the arch and the weight of solids in the arch. In practice this is the case and the stress developed in the arch is related to the size of the hopper outlet, $B$, and the bulk density, $\rho_B$, of the material by the relationship

$$\text{minimum outlet dimension, } B = \frac{H(\theta)\sigma_{crit}}{\rho_B g} \tag{8.4}$$

where $H(\theta)$ is a factor determined by the slope of the hopper wall and $g$ is the acceleration due to gravity. An approximate expression for $H(\theta)$ for conical hoppers is

$$H(\theta) = 2.0 + \frac{\theta}{60} \tag{8.5}$$

### 8.3.7  Summary

From the above discussion of the design philosophy for ensuring mass flow from a conical hopper, we see that the following are required:

(1) the relationship between the strength of the powder in the arch, $\sigma_y$ (unconfined yield stress) with the compacting stress acting on the powder, $\sigma_C$.

(2) the variation of hopper flow factor, $ff$ with

    (a) the nature of the powder (characterized by the effective angle of internal friction, $\delta$);

    (b) the nature of the hopper wall (characterized by the angle of wall friction $\Phi_W$);

    (c) the slope of the hopper wall (characterized by $\theta$, the semi-included angle of the conical section, i.e. the angle between the sloping hopper wall and the vertical).

Knowing $\delta$, $\Phi_W$, and $\theta$, the hopper flow factor $ff$ can be fixed. The hopper flow factor is therefore a function both of powder properties and of the hopper properties (geometry and the material of construction of the hopper walls).

Knowing the hopper flow factor and the powder flow function ($\sigma_y$ versus $\sigma_C$) the critical stress in the arch can be determined and the minimum size of outlet found corresponding to this stress.

## 8.4  SHEAR CELL TESTS

The data listed above can be found by performing shear cell tests on the powder.

The Jenike shear cell (Figure 8.8) allows powders to be compacted to any

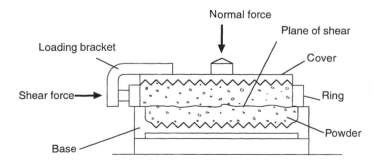

**Figure 8.8**  Jenike shear cell

degree and sheared under controlled load conditions. At the same time the shear force (and hence stress) can be measured.

Generally powders change bulk density under shear. Under the action of shear, for a specific normal load:

- A loosely packed powder would contract (increase bulk density).
- A very tightly packed powder would expand (decrease bulk density).
- A critically packed powder would not change in volume.

For a particular bulk density there is a critical normal load which gives failure (yield) without volume change. A powder flowing in a hopper is in this critical condition. Yield without volume change is therefore of particular interest to us in design.

Using a standardized test procedure five or six samples of powder are prepared all having the same bulk density. Referring to the diagram of the Jenike shear cell shown in Figure 8.8 a normal load is applied to the lid of the cell and the horizontal force applied to the sample via the bracket and loading pin is recorded. That horizontal force necessary to initiate shear of flow of the powder sample is noted. This procedure is repeated for each identical powder sample but with a greater normal load applied to the lid each time. This test thus generates a set of five or six pairs of values for normal load and shear force and hence pairs of values of compacting stress and shear stress for a powder of a particular bulk density. The pairs of values are plotted to give a yield locus (Figure 8.9). The end point of the yield locus corresponds to critical flow conditions where initiation of flow is not accompanied by a change in bulk density. Experience with the procedure permits the operator to select combinations of normal and shear force which achieve the critical conditions. This entire test procedure is repeated two or three times with samples prepared to different bulk densities. In this way a family of yield loci is generated (Figure 8.10).

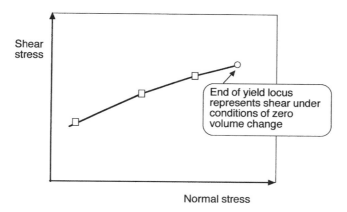

**Figure 8.9**   A single yield locus

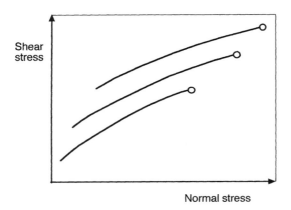

**Figure 8.10**   A family of yield loci

These yield loci characterize the flow properties of the unaerated powder. The following section deals with the generation of the powder flow function from this family of yield loci.

## 8.5   ANALYSIS OF SHEAR CELL TEST RESULTS

The mathematical stress analysis of the flow of unaerated powders in a hopper requires the use of principal stresses. We therefore need to use the Mohr's stress circle in order to determine principal stresses from the results of the shear tests.

### 8.5.1   Mohr's Circle – in Brief

Principal stresses – in any stress system there are two planes at right angles to each other in which the shear stresses are zero. The normal stresses acting on these planes are called the principal stresses.

The Mohr's circle represents the possible combinations of normal and shear stresses acting on any plane in a body (or powder) under stress. Figure 8.11 shows how the Mohr's circle relates to the stress system. Further information on the background to the use of Mohr's circles may be found in most texts dealing with the strength of materials and the analysis of stress and strain in solids.

### 8.5.2   Application of Mohr's Circle to Analysis of the Yield Locus

Each point on a yield locus represents that point on a particular Mohr's circle for which failure or yield of the powder occurs. A yield locus is then tangent to all the Mohr's circles representing stress systems under which the powder will fail (flow). For example, in Figure 8.12 Mohr's circles (a) and (b) represent

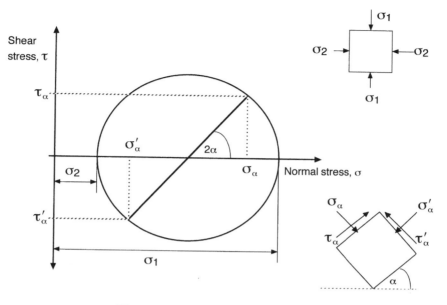

**Figure 8.11** Mohr's circle construction

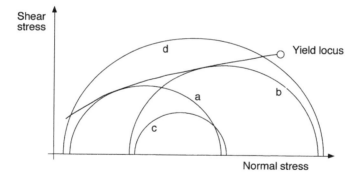

**Figure 8.12** Identification of the applicable Mohr's circle

stress systems under which the powder would fail. In circle (c) the stresses are insufficient to cause flow. Circle (d) this is not relevant since the system under consideration cannot support stress combinations above the yield locus. It is therefore Mohr's circles which are tangential to yield loci that are important to our analysis.

### 8.5.3 Determination of $\sigma_y$ and $\sigma_c$

Two tangential Mohr's circles are of particular interest. Referring to Figure 8.13, the smaller Mohr's circle represents conditions at the free surface of the

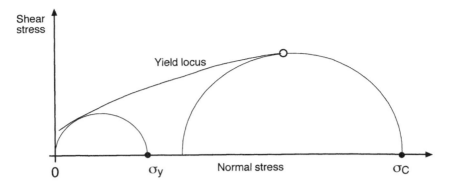

**Figure 8.13**   Determination of unconfined yield stress $\sigma_y$ and compacting stress, $\sigma_C$

arch: this free surface is a plane in which there is zero shear and zero normal stress and so the Mohr's circle which represents flow (failure) under these conditions must pass through the origin of the shear stress versus normal stress plot. This Mohr's circle gives the (major principal) unconfined yield stress, and this is the value we use for $\sigma_y$. The larger Mohr's circle is tangent to the yield locus at its end point and therefore represents conditions for critical failure. The major principal stress from this Mohr's circle is taken as our value of compacting stress, $\sigma_C$.

Pairs of values of $\sigma_y$ and $\sigma_C$ are found from each yield locus and plotted against each other to give the powder flow function (Figure 8.6).

### 8.5.4   Determination of $\delta$ from Shear Cell Tests

Experiments carried out on hundreds of bulk solids (Jenike, 1964) have demonstrated that for an element of powder flowing in a hopper:

$$\frac{\sigma_1}{\sigma_2} = \frac{\text{major principal stress on the element}}{\text{minor principal stress on the element}} = \text{a constant}$$

This property of bulk solids is expressed by the relationship:

$$\frac{\sigma_1}{\sigma_2} = \frac{1 + \sin\delta}{1 - \sin\delta} \tag{8.6}$$

where $\delta$ is the effective angle of internal friction of the solid. In terms of the Mohr's stress circle this means that Mohr's circles for the critical failure are all tangent to a straight line through the origin, the slope of the line being $\tan\delta$ (Figure 8.14).

This straight line is called the effective yield locus of the powder. By drawing in this line, $\delta$ can be determined. Note that $\delta$ is not a real physical angle within the powder; it is the tangent of the ratio of of shear stress to

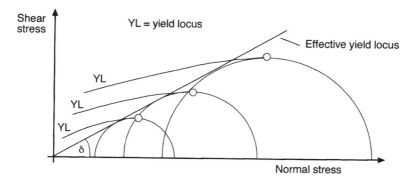

**Figure 8.14** Definition of effective yield locus and effective angle of internal friction, $\delta$

normal stress. Note also that for a free-flowing solid, which does not gain strength under compaction, there is only one yield locus and this locus coincides with the Effective Yield Locus – see Figure 8.15. (This type of relationship between normal stress and shear stress is known as Coulomb friction).

### 8.5.5 The Kinematic Angle of Friction between Powder and Hopper Wall $\Phi_w$

The kinematic angle of friction between powder and hopper wall is otherwise known as the angle of wall friction. This gives us the relationship between normal stress acting between powder and wall and the shear stress under flow conditions. To determine $\Phi_w$ it is necessary to first construct the wall yield locus from shear cell tests. The wall yield locus is determined by shearing the powder against a sample of the wall material under various normal loads. The

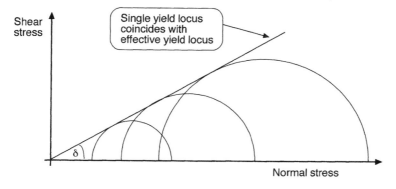

**Figure 8.15** Yield locus for a free-flowing powder

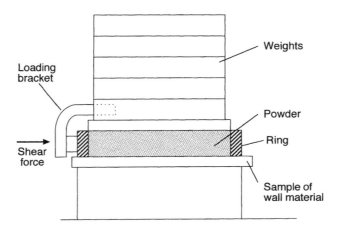

**Figure 8.16**   Apparatus for the measurement of kinematic angle of wall friction, $\Phi_w$

apparatus used is shown in Figure 8.16, and a typical wall yield locus is shown in Figure 8.17.

The kinematic angle of wall friction is given by the gradient of the wall yield locus (Figure 8.17), i.e.

$$\tan \Phi_w = \frac{\text{shear stress at the wall}}{\text{normal stress at the wall}}$$

### 8.5.6   Determination of the Hopper Flow Factor *ff*

The hopper flow factor, *ff*, is a function of $\delta$, $\Phi_w$, and $\theta$ and can be calculated from first principles. However, Jenike (1964) obtained values for a conical hopper and for a wedge-shaped hopper with a slot outlet for values of $\delta$ of 30°, 40°, 50°, 60° and 70°. Examples of the 'flow factor charts' for conical hoppers are shown in Figure 8.18. It will be noticed that values of flow factor exist only

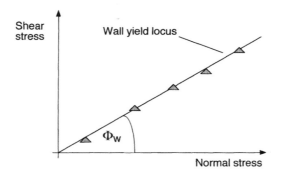

**Figure 8.17**   Kinematic angle of wall friction, $\Phi_w$

in a triangular region; this defines the conditions under which mass flow is possible.

The following is an example of the use of these flow factor charts. Suppose that shear cell tests have given us $\delta$ and $\Phi_w$ equal to 30° and 19° respectively, then entering the chart for conical hoppers with effective angle of friction

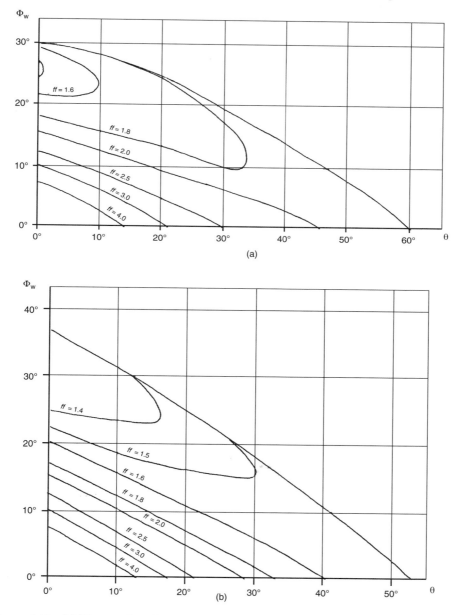

**Figure 8.18**    (a) Hopper flow factor values for conical channels, $\delta = 30°$ (b) Hopper flow factor values for conical channels, $\delta = 40°$

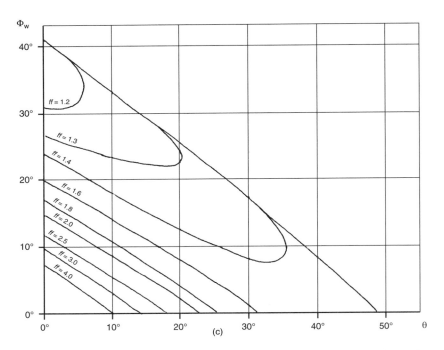

**Figure 8.18**   (c) Hopper flow factor values for conical channels, $\delta = 50°$

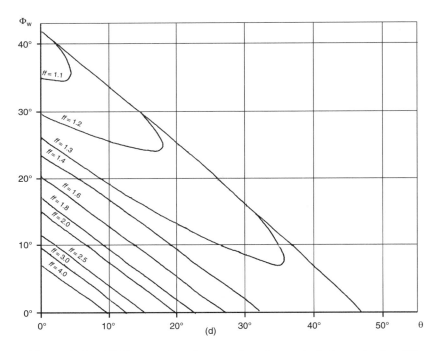

**Figure 8.18**   (d) Hopper flow factor values for conical channels, $\delta = 60°$

$\delta = 30°$, we find that the limiting value of wall slope, $\theta$, to ensure mass flow is 30.5° (point X in Figure 8.19). In practice it is usual to allow a safety margin of 3°, and so, in this case the semi-included angle of the conical hopper $\theta$ would be chosen as 27.5°, giving a hopper flow factor of $ff = 1.8$ (point Y, Figure 8.19).

## 8.6   SUMMARY OF DESIGN PROCEDURE

The following is a summary of the procedure for the design of conical hoppers for mass flows:

(i)   Shear cell tests on powder give a family of yield loci.

(ii)   Mohr's Circle stress analysis gives pairs of values of unconfined yield stress, $\sigma_y$ and compacting stress, $\sigma_C$ and the value of the effective angle of internal friction, $\delta$.

(iii)   Pairs of values of $\sigma_y$ and $\sigma_C$ give the Powder Flow Function.

(iv)   Shear cell tests on the powder and the material of the hopper wall give the Kinematic Angle of Wall Friction, $\Phi_w$.

(v)   $\Phi_w$ and $\delta$ are used to obtain Hopper Flow Factor, $ff$ and semi-included angle of conical hopper wall slope, $\theta$.

(vi)   Powder Flow Function and Hopper Flow Factor are combined to give the stress corresponding to the critical flow – no flow condition, $\sigma_{crit}$.

(vii)   $\sigma_{crit}$, $H(\theta)$ and bulk density, $\rho_B$ are used to calculate the minimum diameter of the conical hopper outlet B.

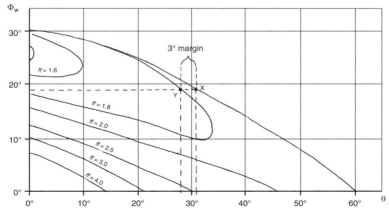

**Figure 8.19**   Worked example of the use of hopper flow factor charts. Hopper flow factor values for conical channels, $\delta = 30°$

## 8.7 DISCHARGE AIDS

A range of devices designed to facilitate flow of powders from silos and hoppers are commercially available. These are known as discharge aids or silo activators. These should not, however, be employed as an alternative to good hopper design.

Discharge aids may be used where proper design recommends an unacceptably large hopper outlet incompatible with the device immediately downstream. In this case the hopper should be designed to deliver uninterrupted mass flow to the inlet of the discharge aid, i.e. the slope of the hopper wall and inlet dimensions of the discharge aid are those calculated according to the procedure outlined in this chapter.

## 8.8 PRESSURE ON THE BASE OF A TALL CYLINDRICAL BIN

It is interesting to examine the variation of stress exerted on the base of a bin with increasing depth of powder. For simplicity we will assume that the powder is non-cohesive (i.e. does not gain strength on compaction). Referring to Figure 8.20, consider a slice of thickness $\Delta H$ at a depth $H$ below the surface of the powder. The downward force is

$$\frac{\pi D^2}{4} \sigma_v \tag{8.7}$$

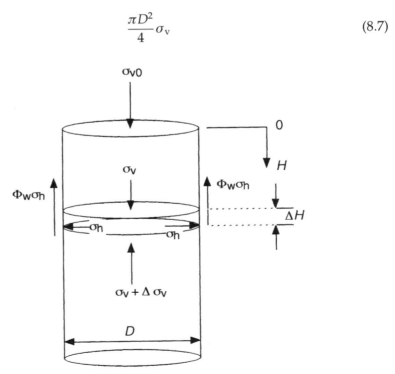

**Figure 8.20**  Forces acting on a horizontal slice of powder in a tall cylinder

where $D$ is the bin diameter and $\sigma_v$ is the stress acting on the top surface of the slice. Assuming stress increases with depth, the reaction of the powder below the slice acts upwards and is

$$\frac{\pi D^2}{4}(\sigma_v + \Delta\sigma_v) \tag{8.8}$$

The net upward force on the slice is then

$$\frac{\pi D^2}{4}\Delta\sigma_v \tag{8.9}$$

If the stress exerted on the wall by the powder in the slice is $\sigma_h$ and the wall friction is $\tan \Phi_w$, then the friction force (upwards) on the slice is

$$\pi D \Delta H \tan \Phi_w \sigma_h \tag{8.10}$$

The gravitational force on the slice is

$$\frac{\pi D^2}{4}\rho_B g \Delta H \quad \text{acting downwards} \tag{8.11}$$

where $\rho_B$ is the bulk density of the powder, assumed to be constant throughout the powder (independent of depth).

If the slice is in equilibrium the upward and downward forces are equated, giving

$$D\Delta\sigma_v + 4\tan \Phi_w \sigma_h \Delta H = D\rho_B g \Delta H \tag{8.12}$$

If we assume that the horizontal stress is proportional to the vertical stress and that the relationship does not vary with depth,

$$\sigma_h = k\sigma_v \tag{8.13}$$

and so as $\Delta H$ tends to zero,

$$\frac{d\sigma_v}{dH} + \left(\frac{4\tan \Phi_w k}{D}\right)\sigma_v = \rho_B g \tag{8.14}$$

Noting that this is the same as

$$\frac{d\sigma_v}{dH}\left[e^{\{(4\tan\Phi_w k/D)\}H}\sigma_v\right] = \rho_B g e^{\{(4\tan\Phi_w k/D)\}H} \tag{8.15}$$

and integrating, we have

$$\sigma_v e^{\{(4\tan \Phi_w k/D)\}H} = \frac{D\rho_B g}{4\tan \Phi_w k}e^{\{(4\tan\Phi_w k/D)\}H} + \text{constant} \tag{8.16}$$

If, in general the stress acting on the surface of the powder is $\sigma_{v0}$ (at $H = 0$) the result is

$$\sigma_v = \frac{D\rho_B g}{4 \tan \Phi_w k}\left[1 - e^{-\{(4\tan\Phi_w k/D)\} H}\right] + \sigma_{v0}e^{-\{(4\tan\Phi_w k/D)\} H} \qquad (8.17)$$

This result was first demonstrated by Janssen (1895).

If there is no force acting on the free surface of the powder, $\sigma_{v0} = 0$ and so

$$\sigma_v = \frac{D\rho_B g}{4 \tan \Phi_w k}\left[1 - e^{-\{(4\tan\Phi_w k/D)\} H}\right] \qquad (8.18)$$

When $H$ is very small

$$\sigma_v \cong \rho_B H g \qquad (8.19)$$

(since for very small $z$, $e^{-z} \cong 1 - z$)

equivalent to the static pressure at a depth $H$ in fluid of density $\rho_B$.

When $H$ is large, inspection of Equation (8.18) gives

$$\sigma_v \cong \frac{D\rho_B g}{4 \tan \Phi_w k} \qquad (8.20)$$

and so the vertical stress developed becomes independent of depth of powder above. The variation in stress with depth of powder for the case of no force acting on the free surface of the powder ($\sigma_{v0} = 0$) is shown in Figure 8.21.

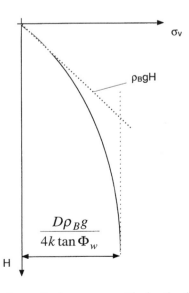

**Figure 8.21**   Variation in vertical pressure with depth of powder (for $\sigma_{v0} = 0$)

Thus, contrary to intuition (which usually based on our experience with fluids), the force exerted by a bed of powder becomes independent of depth if the bed is deep enough. Hence most of the weight of the powder is supported by the walls of the bin. In practice, the stress becomes independent of depth (and also independent of any load applied to the powder surface) beyond a depth of about $4D$.

## 8.9  MASS FLOW RATES

The rate of discharge of powder from an orifice at the base of a bin is found to be independent of the depth of powder unless the bin is nearly empty. This means that the observation for a static powder that the pressure exerted by the powder is independent of depth for large depths is also true for a dynamic system. It confirms that fluid flow theory cannot be applied to the flow of a powder. For flow through an orifice in the flat-based cylinder, experiment shows that:

mass flow rate, $M_p \propto (B - a)^{2.5}$ for a circular orifice of diameter $B$

where $a$ is a correction factor dependent on particle size. (For example, for solids discharge from conical apertures in flat-based cylinders, Beverloo *et al.* (1961) give $M_p = 0.58\rho_B g^{0.5}(B - kx)^{2.5}$.)

For cohesionless coarse particles free falling over a distance $h$ their velocity, neglecting drag and interaction, will be $u = \sqrt{2gh}$.

If these particles are flowing at a bulk density $\rho_B$ through a circular orifice of diameter $B$, then the theortical mass flow rate will be:

$$M_p = \frac{\pi}{4} \sqrt{2} \rho_B g^{0.5} h^{0.5} B^2$$

The practical observation that flowrate is proportional to $B^{2.5}$ and suggests that, in practice, particles only approach the free fall model when $h$ is the same order as the orifice diameter.

## 8.10  CONCLUSIONS

Within the confines of a single chapter it has been possible only to outline the principles involved in the analysis of the flow of unaerated powders. This has been done by reference to the specific example of the design of conical hoppers for mass flow. Other important considerations in the design of hoppers such as time consolidation effects and determination of the stress acting in the hopper and bin wall have been omitted. These aspects together with the details of shear cell testing procedure are covered in texts specific to the subject. Readers wishing to pursue the analysis of failure (flow) in particulate solids in greater detail may refer to texts on soil mechanics.

## 8.11  WORKED EXAMPLES

### WORKED EXAMPLE 8.1

The results of shear cell tests on a powder are shown in Figure 8W1.1. In addition, it is known that the angle of friction on stainless steel is 19° for this powder, and under flow conditions the bulk density of the powder is 1300 kg/m³. A conical stainless steel hopper is to be designed to hold this powder.
   Determine:

(a)  the effective angle of internal friction;

(b)  the maximum semi-included angle of the conical hopper which will con-fidently give mass flow;

(c)  the minimum diameter of the circular hopper outlet necessary to ensure flow when the outlet slide valve is opened.


### Solution

(a)  From Figure 8W1.1, determine the slope of the effective yield locus (line AB). Slope $= 0.578$.

Hence, the effective angle of internal friction, $\delta = \tan^{-1}(0.578) = 30°$

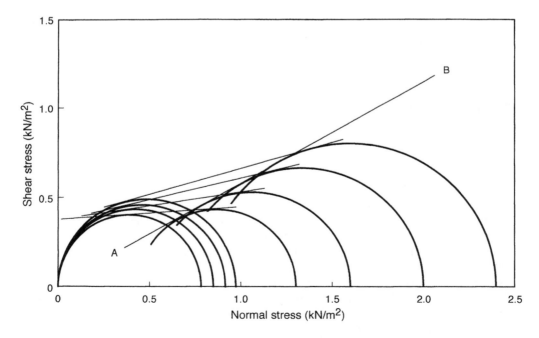

**Figure 8W1.1**   Worked Example 8.1: Shear cell test data

(b) From Figure 8W1.1, determine the pairs of values of $\sigma_C$ and $\sigma_y$ necessary to plot the powder flow function (Figure 8W1.2).

| $\sigma_C$ | 2.4 | 2.0 | 1.6 | 1.3 |
|------------|-----|-----|-----|-----|
| $\sigma_y$ | 0.97 | 0.91 | 0.85 | 0.78 |

Using the flow factor chart for $\delta = 30°$ (Figure 8.18 (a)) with $\Phi_y = 19°$ and a $3°$ margin of safety gives a hopper flow factor, $ff = 1.8$, and the semi-included angle of hopper case, $\theta = 27.5°$ (see Figure 8W1.3).

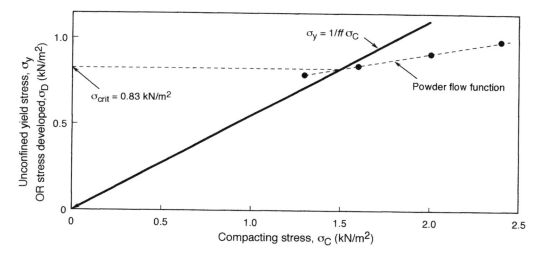

Figure 8W1.2  Worked Example 8.1: Determination of critical stress

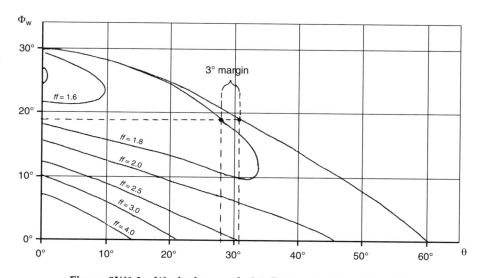

Figure 8W1.3  Worked example 8.1: Determination of $\theta$ and ff

(c)   The relationship $\sigma_y = \sigma_C/ff$ is plotted on the same axes as the powder flow function (Figure 8W1.2) and where this line intercepts the powder flow function we find a value of critical unconfined yield stress, $\sigma_{crit} = 0.83$ kN/m². From Equation (8.5),

$$H(\theta) = 2.46 \text{ when } \theta = 27.5°$$

and from Equation (8.4), the minimum outlet diameter for mass flow, $B$, is

$$B = \frac{2.46 \times 0.83 \times 10^3}{1300 \times 9.81} = 0.160 \text{ m.}$$

Summarizing, then, to achieve mass flow without risk of blockage using the powder in question we require a stainless steel conical hopper with a maximum semi-included angle of cone, 27.5° and a circular outlet with a diameter of at least 16.0 cm.

## WORKED EXAMPLE 8.2

Shear cell tests on a powder give the following information:
Effective angle of internal friction, $\delta = 40°$
Kinematic angle of wall friction on mild steel, $\Phi_w = 16°$
Bulk density under flow condition, $\rho_B = 2000$ kg/m³
The powder flow function which can be represented by the relationship, $\sigma_y = \sigma_C^{0.6}$, where $\sigma_y = $ unconfined yield stress (kN/m²) and $\sigma_C = $ consolidating stress (kN/m²)

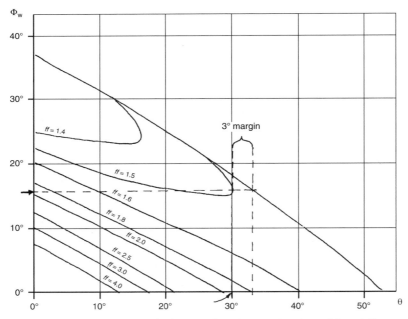

**Figure 8W2.1**   Worked example 2: Determination of $\theta$ and ff

Determine (a) the maximum semi-included angle of a conical mild steel hopper that will confidently ensure mass flow, and (b) the minimum diameter of circular outlet to ensure flow when the outlet is opened.

## Solution

(a) With an effective angle of internal friction $\delta = 40°$ we refer to the flow factor chart in Figure 8.18(b), from which at $\Phi_w = 16°$ and with a safety margin of $3°$ we obtain the hopper flow factor, $ff = 1.5$ and hopper semi-included angle for mass flow, $\theta = 30°$ (see Figure 8W2.1).

(b) For flow: $\dfrac{\sigma_C}{ff} > \sigma_y$ (Equation (8.3))

but for the powder in question $\sigma_y$ and $\sigma_C$ are related by the material flow function:
$\sigma_y = \sigma_C^{0.6}$.
  Thus, the criterion for flow becomes

$$\left[ \frac{\sigma_y^{1/0.6}}{ff} \right] > \sigma_y$$

and so the critical value of unconfined yield stress $\sigma_{\mathrm{crit}}$ is found when $\left[ \dfrac{\sigma_y^{1/0.6}}{ff} \right] = \sigma_y$
hence, $\sigma_{\mathrm{crit}} = 1.837 \ \mathrm{kN/m^2}$.
  From Equation (8.5), $H(\theta) = 2.5$ when $\theta = 30°$ and hence, from Equation (8.4), minimum diameter of circular outlet,

$$B = \frac{2.5 \times 1.837 \times 10^3}{2000 \times 9.81} = 0.234 \ \mathrm{m}.$$

Summarizing, mass flow without blockages is ensured by using a mild steel hopper with maximum semi-included cone angle $30°$ and a circular outlet diameter of at least 23.4 cm.

## EXERCISES

**8.1** Shear cell tests on a powder show that its effective angle of internal friction is $40°$ and its powder flow function can be represented by the equation: $\sigma_y = \sigma_C^{0.45}$, where $\sigma_y$ is the unconfined yield stress and $\sigma_C$ is the compacting stress, both in $\mathrm{kN/m^2}$. The bulk density of the powder is $1000 \ \mathrm{kg/m^3}$ and angle of friction on a mild steel plate is $16°$. It is proposed to store the powder in a mild steel conical hopper of semi-included angle $30°$ and having a circular discharge opening of $0.30 \ \mathrm{m}$ diameter. What is the critical outlet diameter to give mass flow? Will mass flow occur?

(Answer: 0.355 m, no flow)

**8.2** Describe how you would use shear cell tests to determine the effective angle of internal friction of a powder.

A powder has an effective angle of internal friction of 60° and has a powder flow function represented in the graph shown in Figure 8E2.1. If the bulk density of the powder in 1500 kg/m³ and its angle of friction on mild steel plate is 24.5°, determine, for a mild steel hopper, the maximum semi-included angle of cone required to safely ensure mass flow, and the minimum size of circular outlet to ensure flow when the outlet is opened.

(Answer: 17.5°, 18.92 cm)

**8.3**

(a) Summarize the philosophy used in the design of conical hoppers to ensure flow from the outlet when the outlet valve is opened.

(b) Explain how the powder flow function and the effective angle of internal friction are extracted from the results of shear cell tests on a powder.

(c) A firm having serious hopper problems takes on a chemical engineering graduate.

The hopper in question feeds a conveyor belt and periodically blocks at the outlet and needs to be 'encouraged' to restart. The graduate makes an investigation on the

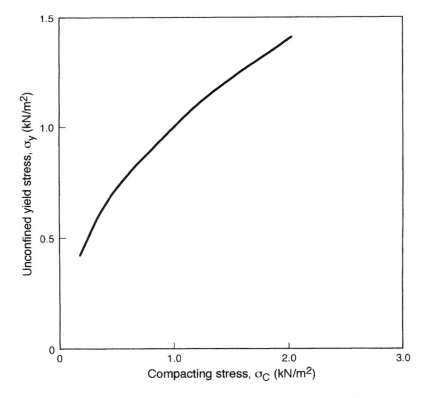

**Figure 8E2.1**   Powder Flow Function for Exercise 8.2

hopper, commissions shear cell tests on the powder and recommends a minor modification to the hopper. After the modification the hopper gives no further trouble and the graduate's reputation is established. Given the information below, what was the graduate's recommendation?

Existing design:      Material of wall – mild steel
Semi-included angle of conical hopper –33°
Outlet – circular, fitted with 25 cm diameter slide valve

Shear cell test data:      Effective angle of internal friction, $\delta = 60°$
Angle of wall friction on mild steel, $\Phi_w = 8°$
Bulk density, $\rho_B = 1250\text{kg/m}^3$
Powder flow function: $\sigma_y = \sigma_C^{0.55}$ ($\sigma_y$ and $\sigma_C$ in kN/m²)

**8.4** Shear cell tests are carried out on a powder for which a stainless steel conical hopper is to be designed. The results of the tests are shown graphically in Figure 8E4.1. In addition it is found that the friction between the powder on stainless steel can be described by an angle of wall friction of 11°, and that the relevant bulk density of the powder is 900 kg/m³.

(a) From the shear cell results of Figure 8E4.1, deduce the effective angle of internal friction $\delta$ of the powder.

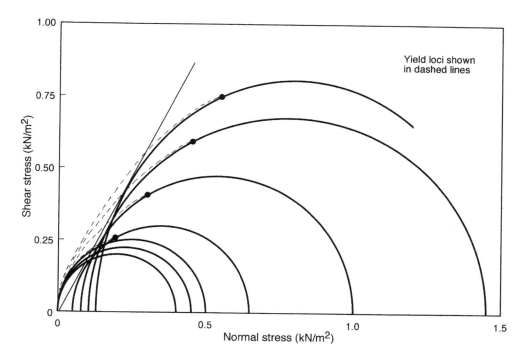

**Figure 8E4.1**    Shear Cell Test Data for Exercise 8.4

(b)  Determine:

    (i)   the semi-included hopper angle safely ensuring mass flow.

    (ii)  the Hopper Flow Factor, $ff$.

    Combine this information with further information gathered from Figure 8E4.1 in order to determine the minimum diameter of outlet to ensure flow when required.

(d)  What do you understand by 'angle of wall friction' and 'effective angle of internal friction'?

(Answer: (a) 60°, (b) (i) 32.5°, (ii) 1.29; (c) 0.110 m – *Note*: extrapolation is necessary here.)

**8.5** The results of shear cell tests on a powder are given in Figure 8E5.1. An aluminium conical hopper is to be designed to suit this powder. It is known that the angle of wall friction between the powder and aluminium is 16° and that the relevant bulk density is 900 kg/m³.

(a)  From Figure 8E5.1 determine the effective angle of internal friction of the powder.

(b)  also determine

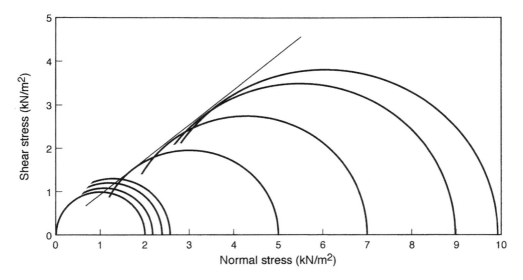

**Figure 8E5.1**   Shear Cell Test Data for Exercise 8.5

(i)   the semi-included hopper angle safely ensuring mass flow.

(ii)   the hopper flow factor, *ff*

(c)   Combine the information with further information gathered from Figure 8E5.1 in order to determine the minimum diameter of circular outlet to ensure flow when required. (*Note*: Extrapolation of these experimental results may be necessary.)

(Answer: (a) 40°; (b)(i) 29.5°, (ii) 1.5; (c) 0.5 m ± approx. 7% depending on the extrapolation).

# 9
# Mixing and Segregation

## 9.1  INTRODUCTION

Achieving good mixing of particulate solids of different size and density is important in many of the process industries, and yet it is not a trivial exercise. For free-flowing powders, the preferred state for particles of different size and density is to remain segregated. This is why in a packet of muesli the large particles come to the top as a result of the vibration caused by handling of the packet. An extreme example of this segregation is that a large steel ball can be made to rise to the top of a beaker of sand by simply shaking the beaker up and down – this has to be seen to be believed! Since the preferred state for free flowing powders is to segregate by size and density, it is not surprising that many processing steps give rise to segregation. Processing steps which promote segregation should not follow steps in which mixing is promoted. In this chapter we will examine mechanisms of segregation and mixing in particulate solids, briefly look at how mixing is carried out in practice and how the quality of a mixture is assessed.

## 9.2  TYPES OF MIXTURE

A perfect mixture of two types of particles is one in which a group of particles taken from any position in the mixture will contain the same proportions of each particle as the proportions present in the whole mixture. In practice, a perfect mixture cannot be obtained. Generally, the aim is to produce a random mixture, i.e. a mixture in which the probability of finding a particle of any component is the same at all locations and equal to the proportion of that component in the mixture as a whole. When attempting to mix particles which are not subject to segregation, this is generally the best quality of mixture that can be achieved. If the particles to be mixed differ in physical properties then segregation may occur. In this case particles of one component have a greater probability of being found in one part of the mixture and so a random mixture cannot be achieved. In Figure 9.1 examples are given of what is meant by

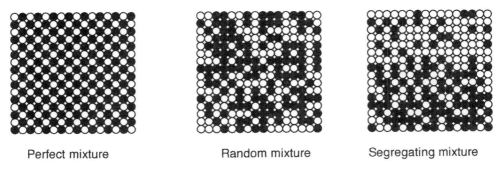

| Perfect mixture | Random mixture | Segregating mixture |

**Figure 9.1**  Types of mixture

perfect, random and segregating mixtures of two components. The random mixture was obtained by tossing a coin – heads gives a black particle at a given location and tails gives a white particle. For the segregating mixture the coin is replaced by a die. In this case the black particles differ in some property which causes them to have a greater probability of appearing in the lower half of the box. In this case, in the lower half of the mixture there is a chance of two in three that a particle will be black (i.e. a throw of 1, 2, 3 or 4) where as in the upper half the probability is one in three (a throw of 5 or 6). It is possible to produce mixtures with better than random quality by taking advantage of the natural attractive forces between particles; such mixtures are achieved through 'ordered' or 'interactive' mixing (see below).

## 9.3   SEGREGATION

### 9.3.1   Causes and Consequences of Segregation

When particles to be mixed have the same important physical properties (size distribution, shape, density) then, provided the mixing process goes on for long enough, a random mixture will be obtained. However, in many common systems, the particles to be mixed have different properties and tend to exhibit segregation. Particles with the same physical property then collect together in one part of the mixture and the random mixture is not a natural state for such a system of particles. Even if particles are originally mixed by some means, they will tend to unmix on handling (moving, pouring, conveying, processing).

Although differences in size, density and shape of the constituent particles of a mixture may give rise to segregation, difference in particle size is by far the most important of these. Density difference is comparatively unimportant (see steel ball in sand example – below) except in gas fluidization where density difference is more important than size difference. Many industrial problems arise from segregation. Even if satisfactory mixing of constituents is achieved in a powder mixing device, unless great care is taken, subsequent processing and handling of the mixture will result in demixing or segregation.

This can give rise to variations in bulk density of the powder going to packaging (e.g. not possible to fit 25 kg into a 25 kg bag) or, more seriously, the chemical composition of the product may be off specification (e.g. in blending of constituents for detergents or drugs).

## 9.3.2 Mechanisms of Segregation

Four mechanisms (Williams, 1990) of segregation according to size may be identified:

(1) *Trajectory segregation.* If a small particle of diameter $x$ and density $\rho_p$, whose drag is governed by Stokes' law, is projected horizontally with a velocity $U$ into a fluid of viscosity $\mu$ and density $\rho_f$, the limiting distance that it can travel horizontally is $U\rho_p x^2/18\mu$.

---

From Chapter 1, the retarding force on the particle $= C_D \frac{1}{2}\rho_f U^2 \left(\dfrac{\pi x^2}{4}\right)$

Deceleration of the particle $= \dfrac{\text{retarding force}}{\text{mass of particle}}$

In Stokes' law region, $C_D = 24/Re_p$

Hence, deceleration $= \dfrac{18U\mu}{\rho_p x^2}$

From the equation of motion a particle with an initial velocity $U$ and constant deceleration $18U\mu/\rho_p x^2$ will travel a distance $U\rho_p x^2/18\mu$ before coming to rest.

---

A particle of diameter $2x$ would therefore travel four times as far before coming to rest. This mechanism can cause segregation where particles are caused to move through the air (Figure 9.2). This also happens when powders fall from the end of a conveyor belt.

(2) *Percolation of fine particles.* If a mass of particles is disturbed in such a way that individual particles move, a rearrangement in the packing of the particles occurs. The gaps created allow particles from above to fall, and particles in some other place to move upwards. If the powder is composed of particles of different size, it will be easier for small particles to fall down and so there will be a tendency for small particles to move downwards leading to segregation. Even a very small difference in particle size can give rise to significant segregation.

Segregation by percolation of fine particles can occur whenever the mixture is disturbed, causing rearrangement of particles. This can happen during stirring, shaking, vibration or when pouring particles into a heap. Note that

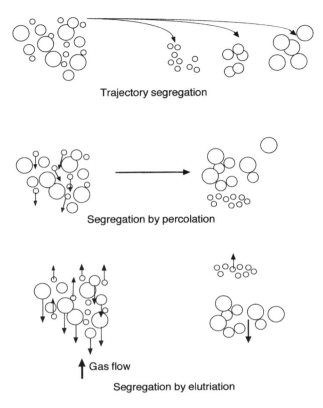

Trajectory segregation

Segregation by percolation

Gas flow

Segregation by elutriation

**Figure 9.2**   Mechanisms of segregation

stirring, shaking and vibration would all be expected to promote mixing in liquids or gases, but cause segregation in free-flowing particle mixtures. Figure 9.3 shows segregation in the heap formed by pouring a mixture of two sizes of particles. The shearing caused when a particle mixture is rotated in a drum can also give rise to segregation by percolation.

Segregation by percolation occurs in charging and discharging storage hoppers. As particles are fed into a hopper they generally pour into a heap resulting in segregation if there is a size distribution and the powder is free-flowing. There are some devices and procedures available to minimize this effect if segregation is a particular concern. However, during discharge of a core flow hopper (see Chapter 8) sloping surfaces form, along which particles roll, and this gives rise to segregation in free-flowing powders. If segregation is a cause for concern, therefore, core flow hoppers should be avoided.

(3) *Rise of coarse particles on vibration*. If a mixture of particles of different sizes is vibrated the larger particles move upwards. This can be demonstrated by placing a single large ball at the bottom of a bed of sand (for example, a 20 mm steel ball or similarly sized pebble in a beaker of sand from the beach). On shaking the beaker up and down, the steel ball rises to the surface. Generally

**Figure 9.3**  Segregation pattern formed by pouring a free-flowing mixture of two sizes of particles into a heap

the segregation effect increases with increasing density of the large particle. The mechanism responsible for this effect is not entirely understood, but two possibilities are given below.

When the system is moving upwards during the shaking motion, the momentum of the large particle causes it to penetrate the loosely packed fine particles. When the system is moving downwards the large particle causes an increase in pressure in the region below it, which compacts the fine particles, creating resistance to the downward motion of the large particle. With each cycle of the vibration the large particle moves up 'one notch'. Figure 9.4 show a series of photographs taken from a 'two dimensional' version of the steel ball experiment. Another possible mechanism for vibrational segregation is proposed by Knight et al. (1993), who suggest that the action of the container walls on the fine particles sets up convection patterns with the fine particles rising in the centre and falling near to the walls. The convective motion of the fine particles lifts the larger particle to the top of the powder mass.

(4) *Elutriation segregation.* When a powder containing an appreciable proportion of particles under 50 μm is charged into a storage vessel or hopper, air is displaced upwards. The upward velocity of this air may exceed the terminal free-fall velocity (see Chapter 1) of some of the finer particles, which may then remain in suspension after the larger particles have settled to the surface of the hopper contents (Figure 9.2). For particles in this size range in air the terminal free-fall velocity will be typically of the order of a few centimetres per second and will increase as the square of particle diameter (e.g. for 30 μm sand particles the terminal velocity is 7 cm/s). Thus a pocket of fine particles is generated in the hopper each time solids are charged.

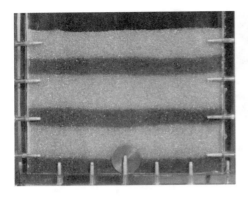
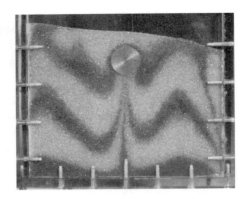

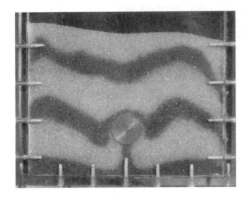
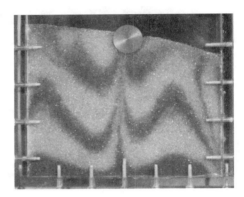

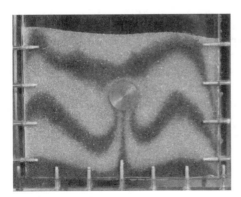

**Figure 9.4** Series of photographs showing the rise of a steel disc through a bed of 2 mm glass spheres due to vibration (a 'two dimensional' version of the rising steel ball experiment)

## 9.4 REDUCTION OF SEGREGATION

Segregation occurs primarily as a result of size difference. The difficulty of mixing two components can therefore be reduced by making the size of the components as similar as possible and by reducing the absolute size of both components. Segregation is generally not a serious problem when all particles are less than 30 μm (for particle densities in the range 2000–3000 kg/m³). In such fine powders the interparticle forces (see Chapter 11) generated by electrostatic charging, van der Waals forces and forces due to moisture are large compared with the gravitational and inertial forces on the particles. This causes the particles to stick together preventing segregation as the particles are not free to move relative to one another. These powders are referred to as cohesive powders (Geldart's classification of powders for fluidization is relevant here – see Chapter 5). The lack of mobility of individual particles in cohesive powders is one reason why they give better quality of mixing. The other reason is that if a random mixture is approached, the standard deviation of the composition of samples taken from the mixture will decrease in inverse proportion to the number of particles in the sample. Therefore, for a given mass of sample the standard deviation decreases and mixture quality increases with decreasing particle size. The mobility of particles in free-flowing powders can be reduced by the addition of small quantities of liquid. The reduction in mobility reduces segregation and permits better mixing.

It is possible to take advantage of this natural tendency for particles to adhere to produce mixtures of quality better that random mixtures. Such mixtures are known as ordered or interactive mixtures; they are made up of small particles (e.g. < 5 μm) adhered to the surface of a carrier particle in a controlled manner (Figure 9.5). By careful selection of particle size and engineering of interparticle forces, high quality mixtures with very small variance can be achieved. This technique is use in the pharmaceutical industry where quality control standards are exacting. For further details on ordered mixing and on the mixing of cohesive powders the reader is referred to Harnby *et al.* (1992).

If it is not possible to alter the size of the components of the mixture or to add liquid, then in order to avoid serious segregation, care should be taken to avoid situations which are likely to promote segregation. In particular pouring

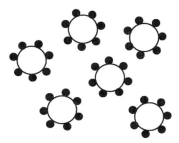

**Figure 9.5**   An ordered mixture of small particles on carrier particles

operations and the formation of a moving sloping powder surface should be avoided.

## 9.5   EQUIPMENT FOR PARTICULATE MIXING

### 9.5.1   Mechanisms of Mixing

Lacey (1954) identified three mechanisms of powder mixing:

(1)   convective mixing;

(2)   diffusive mixing;

(3)   shear mixing.

In shear mixing, shear stresses give rise to slip zones and mixing takes place by interchange of particles between layers within the zone. Diffusive mixing occurs when particles roll down a sloping surface. Convective mixing is by deliberate bulk movement of packets of powder around the powder mass.

In free-flowing powders both diffusive mixing and shear mixing give rise to size segregation and so for such powders convective mixing is the major mechanism promoting mixing.

### 9.5.2   Types of Mixer

(1)   *Tumbling mixers.* A tumbling mixer comprises a closed vessel rotating about its axis. Common shapes for the vessel are cube, double cone and V (Figure 9.6). The dominant mechanism is diffusive mixing. Since this can give rise to segregation in free-flowing powders the quality of mixture achievable with such powder in tumbling mixers is limited. Baffles may be installed in an attempt to reduce segregation, but have little effect.

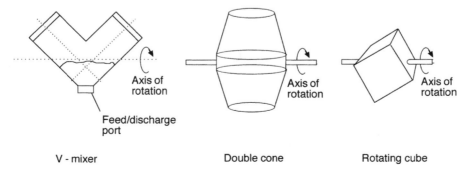

**Figure 9.6**   Tumbling mixers. V-mixer, double cone mixer and rotating cube mixer

(2)  *Convective mixers.* In convective mixers circulation patterns are set up within a static shell by rotating blades or paddles. The main mechanism is convective mixing as the name suggests, although this is accompanied by some diffusive and shear mixing. One of the most common convective mixers is the ribbon blender in which helical blades or ribbons rotate on a horizontal axis in a static cylinder or trough (Figure 9.7). Rotational speed are typically less than one revolution per second. A somewhat different type of convective mixer is the Nautamix (Figure 9.8) in which an Archimedean screw lifts material from the base of a conical hopper and progresses around the hopper wall.

(3)  *Fluidized bed mixers.* These rely on the natural mobility afforded particles in the fluidized bed. The mixing is largely convective with the circulation patterns set up by the bubble motion within the bed. An important feature of the fluidized bed mixer is that several processing steps (e.g.

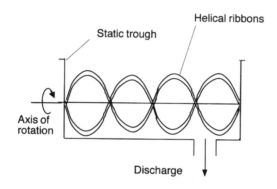

**Figure 9.7**   Ribbon blender

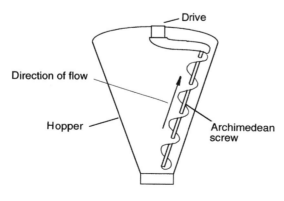

**Figure 9.8**   Schematic diagram of the Nautamixer

mixing, reaction, coating drying etc.) may be carried out in the same vessel.

(4)  *High shear mixers*. Local high shear stresses are created by devices similar to those used in comminution; for example high velocity rotating blades, low velocity–high compression rollers (see Chapter 10). In the high shear mixers the emphasis is on breaking down agglomerates of cohesive powders rather than breaking individual particles. The dominant mechanism is shear mixing.

## 9.6  ASSESSING THE MIXTURE

### 9.6.1  Quality of a Mixture

The end use of a particle mixture will determine the quality of mixture required. The end use imposes a scale of scrutiny on the mixture. 'Scale of scrutiny' was a term used by Danckwerts (1953) meaning 'the maximum size of the regions of segregation in the mixture which would cause it to be regarded as imperfectly mixed'. For example, the appropriate scale of scrutiny for a detergent powder composed of active ingredients in particulate form is the quantity of detergent in the scoop used to dipense it into the washing machine. The composition should not vary significantly between the first and last scoops taken from the box. At another extreme the scale of scrutiny for a pharmaceutical drug is the quantity of material making up the tablet or capsule. The quality of a mixture decreases with decreasing scale of scrutiny until in the extreme we are scrutinizing only individual particles. An example of this is the image on a television screen, which at normal viewing distance appears as a life-like image, but which under close 'scrutiny' is made up of tiny dots of red, green and blue colour.

### 9.6.2  Sampling

To determine the quality of a mixture it is generally necessary to take samples. In order to avoid bias in taking samples from a particulate mixture, the guidelines of sampling powders set out in Chapter 1 must be followed. The size of the sample required in order to determine the quality of the mixture is governed by the scale of scrutiny imposed by the intended use of the mixture.

### 9.6.3  Statistics Relevant to Mixing

It is evident that the sampling of mixtures and the analysis of mixture quality require the application of statistical methods. The statistics relevant to random binary mixtures are summarized below:

- *Mean composition.* The true composition of a mixture $\mu$ is often not known but an estimate $\overline{y}$ may be found by sampling. If we have $N$ samples of composition $y_1$ to $y_N$ in one component, the estimate of the mixture composition $\overline{y}$ is given by

$$\overline{y} = \frac{1}{N} \sum_{i=1}^{N} y_i \tag{9.1}$$

- *Standard deviation and variance.* The true standard deviation $\sigma$ and the true variance, $\sigma^2$ of the composition of the mixture are quantitative measures of the quality of the mixture. The true variance is usually not known but an estimate $S^2$ is defined as

$$S^2 = \frac{\sum_{i=1}^{N}(y_i - \mu)^2}{N} \qquad \text{if the true composition } \mu \text{ is known} \tag{9.2}$$

$$S^2 = \frac{\sum_{i=1}^{N}(y_i - \overline{y})^2}{N-1} \qquad \text{if the true composition } \mu \text{ is unknown} \tag{9.3}$$

The standard deviation is equal to the square root of variance.

- *Theoretical limits of variance.* For a two-component system the theoretical upper and lower limits of mixture variance are:

(a)   upper limit (completely segregated)   $\sigma_0^2 = p(1 - p)$ $\tag{9.4}$

(b)   lower limit (randomly mixed)   $\sigma_R^2 = \dfrac{p(1 - p)}{n}$ $\tag{9.5}$

where $p$ and $(1 - p)$ are the proportions of the two components determined from samples and $n$ is the number of particles in each sample.

Actual values of mixture variance lie between these two extreme values.

- *Mixing indices.* A measure of the degree of mixing is the Lacey mixing index (Lacey, 1954):

$$\text{Lacey mixing index} = \frac{\sigma_0^2 - \sigma^2}{\sigma_0^2 - \sigma_R^2} \tag{9.6}$$

In practical terms the Lacey mixing index is the ratio of 'mixing achieved' to 'mixing possible'. A Lacey mixing index of zero would represent complete segregation and a value of unity would represent a completely random mixture. Practical values of this mixing index, however, are found to lie in the range 0.75 to 1.0 and so the Lacey mixing index does not provide sufficient discrimination between mixtures.

A further mixing index suggested by Poole *et al.* (1964) is defined as:

$$\text{Mixing index of Poole } et\ al.: = \frac{\sigma}{\sigma_R} \tag{9.7}$$

This index gives better discrimination for practical mixtures and approaches unity for completely random mixtures.

- *Standard error*. When the sample compositions have a normal distribution the sampled variance values will also have a normal distribution. The standard deviation of the variance of the sample compositions is known as the 'standard error' of the variance $E(S^2)$.

- *Tests for precision of mixture composition and variance*. The mean mixture composition and variance which we measure from sampling is only a sample from the normal distribution of mixture compositions and variance values for that mixture. We need to be able to assign a certain confidence to this estimate and to determine its precision.
    Assuming that the sample compositions are normally distributed,

(1) *Sample composition*
    Based on $N$ samples of mixture composition with mean $\bar{y}$ and estimated standard deviation $S$, the true mixture composition $\mu$ may be stated with precision:

$$\mu = \bar{y} \pm \frac{tS}{\sqrt{N}} \tag{9.8}$$

   where $t$ is from Student's $t$ test for statistical significance. The value of $t$ depends on the confidence level required. For example, at 95% confidence level, $t = 2.0$ for $N = 60$, and so there is a 95% probability that the true mean mixture composition lies in the range: $\bar{y} \pm 0.258S$. In other words, 1 in 20 estimates of mixture variance estimates would lie outside this range.

(2) *Variance*

   (a) When more than 50 samples are taken (i.e. $n > 50$), the distribution of variance values can also be assumed to be normal and the Student's $t$ test may be used. The best estimate of the true variance $\sigma^2$ is then given by

$$\sigma^2 = S^2 \pm \{t \times E(S^2)\} \tag{9.9}$$

The standard error of the mixture variance required in this test is usually not known but is estimated from:

$$E(S^2) = S^2 \sqrt{\frac{2}{N}}$$
(9.10)

The standard error decreases as $1/\sqrt{N}$ and so the precision increases as $\sqrt{N}$

(b) When less than 50 samples are taken (i.e. $n < 50$), the variance distribution curve may not be normal and is likely to be a $\chi^2$ (chi squared) distribution. In this case the limits of precision are not symmetrical. The range of values of mixture variance is defined by lower and upper limits:

$$\text{lower limit:} \quad \sigma_L^2 = \frac{S^2(N-1)}{\chi_\alpha^2}$$
(9.11)

$$\text{upper limit:} \quad \sigma_U^2 = \frac{S^2(N-1)}{\chi_{1-\alpha}^2}$$
(9.12)

Where $\alpha$ is the significance level (for a 90% confidence range, $\alpha = 0.5(1 - 90/100) = 0.05$; for a 95% confidence range, $\alpha = 0.5(1 - 95/100) = 0.025$). The lower and upper $\chi^2$ values, $\chi_\alpha^2$ and $\chi_{1-\alpha}^2$ for a given confidence level are found in $\chi^2$ distribution tables.
*Note:* Standard statistical tables, which are widely and readily available, are a source of Student's $t$ values and $\chi^2$ (chi squared) distribution values.

## 9.7   WORKED EXAMPLES

### WORKED EXAMPLE 9.1 (AFTER WILLIAMS, 1990)

A random mixture consists of two components A and B in proportions 60% and 40% by mass respectively. The particles are spherical and A and B have particle densities 500 and 700 kg/m³ respectively. The cumulative undersize mass distributions of the two components are shown in Table 9W1.1.

**Table 9W1.1**   Size distributions of particles A and B for Worked Example 9.1

| Size $x$ (μm) | 2057 | 1676 | 1405 | 1204 | 1003 | 853 | 699 | 599 | 500 | 422 |
|---|---|---|---|---|---|---|---|---|---|---|
| $F_A(x)$ | 1.00 | 0.80 | 0.50 | 0.32 | 0.19 | 0.12 | 0.07 | 0.04 | 0.02 | 0 |
| $F_B(x)$ | | | 1.00 | 0.88 | 0.68 | 0.44 | 0.21 | 0.08 | 0 | |

If samples of 1 g are withdrawn from the mixture, what is the expected value for the standard deviation of the composition of the samples?

## Solution

The first step is to estimate the number of particles per unit mass of A and B. This is done by converting the size distributions into differential frequency number distributions and using:

$$\begin{bmatrix} \text{mass of particles} \\ \text{in each size range} \end{bmatrix} = \begin{bmatrix} \text{number of particles} \\ \text{in size range} \end{bmatrix} \times \begin{bmatrix} \text{mass of one} \\ \text{particle} \end{bmatrix}$$

$$\mathrm{d}m \qquad = \qquad \mathrm{d}n \qquad\qquad \frac{\rho_p \pi x^3}{6}$$

where, $\rho_p$ is the particle density and $x$ is the arithmetic mean of adjacent sieve sizes.

These calculations are summarised in Tables 9W1.2 and 9W1.3.

Table 9W1.2    'A' particles (Worked Example 9.1)

| Mean size of range $x$ (μm) | $\mathrm{d}m$ | $\mathrm{d}n$ |
|---|---|---|
| 1866.5 | 0.20 | 117 468 |
| 1540.5 | 0.30 | 334 081 |
| 1304.5 | 0.18 | 309 681 |
| 1103.5 | 0.13 | 369 489 |
| 928 | 0.07 | 334 525 |
| 776 | 0.05 | 408 658 |
| 649 | 0.03 | 419 143 |
| 549.5 | 0.02 | 460 365 |
| 461 | 0.02 | 779 655 |
| Totals | 1.00 | $3.51 \times 10^6$ |

Table 9W1.3    'B' particles (worked example 9.1)

| Mean size of range $x$ (μm) | $\mathrm{d}m$ | $\mathrm{d}n$ |
|---|---|---|
| 1866.5 | 0 | 0 |
| 1540.5 | 0 | 0 |
| 1304.5 | 0.12 | $0.147 \times 10^6$ |
| 1103.5 | 0.20 | $0.406 \times 10^6$ |
| 928 | 0.24 | $0.819 \times 10^6$ |
| 776 | 0.23 | $1.343 \times 10^6$ |
| 649 | 0.13 | $1.297 \times 10^6$ |
| 549.5 | 0.08 | $1.315 \times 10^6$ |
| 461 | 0 | 0 |
| Totals | 1.00 | $5.33 \times 10^6$ |

Thus, $n_A = 3.51 \times 10^6$ particles per kg.

and, $n_B = 5.33 \times 10^6$ particles per kg.

And in samples of 1 g (0.001 kg) we would expect a total number of particles:

$$n = 0.001 \times (3.51 \times 10^6 \times 0.6 + 5.33 \times 10^6 \times 0.4)$$

$$= 4238 \text{ particles}$$

And so, from Equation (9.5) for a random mixture,

$$\text{standard deviation, } \sigma = \sqrt{\frac{0.6 \times 0.4}{4238}} = 0.0075$$

## WORKED EXAMPLE 9.2 (AFTER WILLIAMS, 1990)

Sixteen samples are removed from a binary mixture and the percentage proportions of one component by mass are:

$$41, 37, 41, 39, 45, 37, 39, 40$$

$$41, 43, 40, 38, 39, 37, 43, 40$$

Determine the upper and lower 95% and 90% confidence limits for the standard deviation of the mixture.

### Solution to Worked Example 9.2:

From Equation (9.1), the mean value of the sample compositions is

$$\overline{y} = \frac{1}{16} \sum_{i=1}^{16} y_i = 40\%$$

Since the true mixture composition is not known an estimate of the standard deviation is found from Equation (9.3):

$$S = \sqrt{\left[\frac{1}{16-1} \sum_{i=1}^{16} (y_i - 40)^2\right]} = 2.31$$

Since there are less than 50 samples, the variance distribution curve is more likely to be a $\chi^2$ distribution. Therefore, from Equations (9.11) and (9.12):

$$\text{lower limit:} \quad \sigma_L^2 = \frac{2.31^2(16-1)}{\chi_\alpha^2}$$

$$\text{upper limit:} \quad \sigma_U^2 = \frac{2.31^2(16-1)}{\chi_{1-\alpha}^2}$$

**Table 9W3.1**   Size distributions of drug and exipient (Worked Example 9.3)

| Size $x$ (µm) | 420 | 355 | 250 | 190 | 150 | 75 | 53 | 0 |
|---|---|---|---|---|---|---|---|---|
| $F_D(x)$ | 1.00 | 0.991 | 0.982 | 0.973 | 0.964 | 0.746 | 0.046 | 0 |
| $F_E(x)$ | 1.00 | 1.00 | 0.977 | 0.967 | 0.946 | 0.654 | 0.284 | 0 |

At the 90% confidence level $\alpha = 0.05$ and so referring to the $\chi^2$ distribution tables with 15 degrees of freedom $\chi^2_a = 24.996$ and $\chi^2_{1-a} = 7.261$.
Hence: $\sigma^2_L = 3.2$ and $\sigma^2_U = 11.02$.
At the 95% confidence level $\alpha = 0.025$ and so referring to the $\chi^2$ distribution tables with 15 degrees of freedom $\chi^2_a = 27.49$ and $\chi^2_{1-a} = 6.26$.
Hence: $\sigma^2_L = 2.91$ and $\sigma^2_U = 12.78$.

## WORKED EXAMPLE 9.3

During the mixing of a drug with an excipient the standard deviation of the compositions of 100 milligram samples tends to a constant value of ±0.005. The size distributions of drug (D) and exipient (E) are given in Table 9W3.1.
The mean proportion by mass of drug is know to be 0.2. The densities of drug and excipient are 1100 and 900 kg/m³ respectively.
   Determine whether the mixing is satisfactory (a) if the criterion is a random mixture and (b) if the criterion is an in-house specification that the composition of 95% of the samples should lie within ±15% of the mean.

### Solution

The number of particles of drug and excipient in each sample is first calculated as shown in Worked Example 9.1:

**Table 9E3.2**   Number of drug particles in each kg of sample (Worked Example 9.3)

| Mean size of range $x$ (µm) | $dm$ | $dn$ |
|---|---|---|
| 388 | 0.009 | $2.67 \times 10^5$ |
| 303 | 0.009 | $5.62 \times 10^5$ |
| 220 | 0.009 | $1.47 \times 10^6$ |
| 170 | 0.009 | $3.18 \times 10^6$ |
| 113 | 0.218 | $2.62 \times 10^8$ |
| 64 | 0.700 | $4.64 \times 10^9$ |
| 27 | 0.046 | $4.06 \times 10^9$ |
| 20 | 0.00 | 0 |
| 0 | 0.00 | 0 |
| Totals | 1.00 | $8.96 \times 10^9$ |

**Table 9E3.3** Number of excipient particles in each kg of sample (Worked Example 9.3)

| Mean size of range $x$ (µm) | d$m$ | d$n$ |
|---|---|---|
| 388 | 0 | 0 |
| 303 | 0.023 | $1.75 \times 10^6$ |
| 220 | 0.010 | $1.99 \times 10^6$ |
| 170 | 0.021 | $9.07 \times 10^6$ |
| 113 | 0.292 | $4.29 \times 10^8$ |
| 64 | 0.374 | $3.03 \times 10^9$ |
| 27 | 0.28 | $3.02 \times 10^{10}$ |
| 20 | 0.00 | 0 |
| 0 | 0 | 0 |
| Totals | 1.00 | $3.37 \times 10^{10}$ |

Thus, $n_D = 8.96 \times 10^9$ particles per kg.

and $n_E = 3.37 \times 10^{10}$ particles per kg.

And in samples of 1 g (0.001 kg) we would expect a total number of particles:

$$n = 100 \times 10^{-6} \times (8.96 \times 10^9 \times 0.2 + 3.37 \times 10^{10} \times 0.8)$$

$$= 2.88 \times 10^6 \text{ particles}$$

And so, from Equation (9.5) for a random mixture,

$$\text{standard deviation, } \sigma_R = \sqrt{\frac{0.2 \times 0.8}{2.88 \times 10^6}} = 0.000235$$

*Conclusion.* The actual standard deviation of the mixture is greater than that for a random mixture and so the criterion of random mixing is not achieved.

For a normal distribution the in-house criterion that 95% of samples should lie within ±15% of the mean suggests that:

$$1.96\sigma = 0.15 \times 0.2$$

(since for a normal distribution 95% of the values lie within ±1.96 standard deviations of the mean).

Hence, $\sigma = 0.0153$. So the in-house criterion is achieved.

## EXERCISES

**9.1** 31 samples are removed from a binary mixture and the percentage proportions of one component by mass are:

19, 22, 20, 24, 23, 25, 22, 18, 24, 21, 27, 22, 18, 20, 23, 19,

20, 22, 25, 21, 17, 26, 21, 24, 25, 22, 19, 20, 24, 21, 23

Determine the upper and lower 95% confidence limits for the standard deviation of the mixture.

(Answer: 0.355 to 0.595)

**9.2** A random mixture consists of two components A and B in proportions 30% and 70% by mass respectively. The particles are spherical and components A and B have particle densities 500 and 700 kg/m$^3$ respectively. The cumulative undersize mass distributions of the two components are shown in Table 9E2.1.

If samples of 5 g are withdrawn from the mixture, what is the expected value for the standard deviation of the composition of the samples?

(Answer: 0.0025)

**9.3** During the mixing of a drug with an excipient the standard deviation of the compositions of 10 mg samples tends to a constant value of ±0.005. The size distributions by mass of drug ($D$) and exipient ($E$) are given in Table 9E3.1.
    The mean proportion by mass of drug is known to be 0.1. The densities of the drug and the excipient are 800 and 1000 kg/m$^3$ respectively.
Determine whether the mixing is satisfactory if

(a)    the criterion is a random mixture and

(b)    the criterion is an in-house specification that the composition of 99% of the samples should lie within ±20% of the mean.

(Answer: (a) 0.00118, criterion not achieved; (b) 0.00775, criterion achieved)

**Table 9E3.1**   Size distributions of drug and excipient (question 9.3)

| Size $x$ (μm) | 499 | 420 | 355 | 250 | 190 | 150 | 75 | 53 | 0 |
|---|---|---|---|---|---|---|---|---|---|
| $F_D(x)$ | 1.00 | 0.98 | 0.96 | 0.94 | 0.90 | 0.75 | 0.05 | 0.00 | 0.00 |
| $F_E(x)$ | 1.00 | 1.00 | 0.97 | 0.96 | 0.93 | 0.65 | 0.25 | 0.05 | 0.00 |

**Table 9E2.1**   Size distributions for particles A and B (Exercise 9.2)

| Size $x$ (μm) | 2057 | 1676 | 1405 | 1204 | 1003 | 853 | 699 | 599 | 500 | 422 | 357 |
|---|---|---|---|---|---|---|---|---|---|---|---|
| $F_A(x)$ | 1.00 | 1.00 | 0.85 | 0.55 | 0.38 | 0.25 | 0.15 | 0.10 | 0.07 | 0.02 | 0.00 |
| $F_B(x)$ |  | 1.00 | 0.80 | 0.68 | 0.45 | 0.25 | 0.12 | 0.06 | 0.00 | 0.00 |

# 10
# Particle Size Reduction

## 10.1   INTRODUCTION

Size reduction, or comminution, is an important step in the processing of many solid materials. It may be used to create particles of a certain size and shape, to increase the surface area available for chemical reaction or to liberate valuable minerals held within particles.

The size reduction of solids is an energy intensive and highly inefficient process: 5% of all electricity generated is used in size reduction; based on the energy required for the creation of new surface, the industrial scale process is generally less than 1% efficient. The two statements would indicate that there is great incentive to improve the efficiency of size reduction processes. However, in spite of a considerable research effort over the years, size reduction processes have remained stubbornly inefficient. Also in spite of the existence of a well-developed theory for a strength and breakage mechanism of solids, the design and scale-up of comminution processes is usually based on past experience and testing, and is very much in the hands of the manufacturer of comminution equipment.

The chapter is intended as a introduction to the topic of size reduction covering the concepts and models involved and including a broad survey of practical equipment and systems. The chapter falls into the following section:

- particle fracture mechanisms;

- models for prediction of energy requirements and product size distribution;

- equipment: matching machine to material and duty.

## 10.2   PARTICLE FRACTURE MECHANISMS

Consider a crystal of sodium chloride (common salt) as a simple and convenient model of a brittle material. Such a crystal is composed of a lattice of positively charged sodium ions and negatively charged chloride ions arranged such that each ion is surrounded by six ions of the opposite sign. Between the oppositely charged ions there is an attractive force whose magnitude is inversely proportional to the square of the separation of the ions. There is also a repulsive force between the negatively charged electron clouds of these ions which becomes important at very small interatomic distances. Therefore two oppositely charged ions have an equilibrium separation such that the attractive and repulsive forces between them are equal and opposite. Figure 10.1 shows how the sum of the attractive and repulsive forces varies with changing separation of the ions. It can be appreciated that if the separation of the ions is increased or decreased by a small amount from the equilibrium separation there will be a resultant net force restoring the ions to the equilibrium position. The ions in the sodium chloride crystal lattice are held in equilibrium positions governed by the balance between attractive and repulsive forces. Over a small range of interatomic distances the relationship between applied tensile or compressive force and resulting change in ion separation is linear. That is, in this region (AB in Figure 10.1) Hooke's law applies: strain is directly proportional to applied stress. The Young's modulus of the material (stress/strain) describes this proportionality. In this Hooke's law range the deformation of the crystal is elastic; i.e. the original shape of the crystal is recovered upon removal of the stress.

In order to break the crystal it is necessary to separate adjacent layers of ions in the crystal and this involves increasing the separation of the adjacent ions beyond the region where Hooke's law applies – i.e. beyond point B in Figure 10.1 into the plastic deformation range. The applied stress required to induce this plastic behaviour is known as the elastic limit or yield stress, and is sometimes defined as the material's strength. With a knowledge of the magnitude of attractive and repulsive forces between ions in such a crystal, it

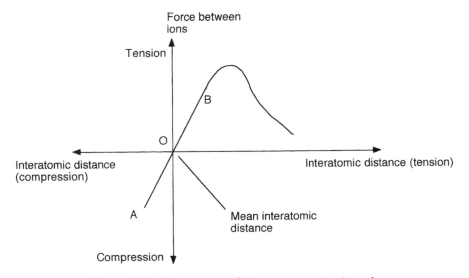

**Figure 10.1**   Force versus distance on an atomic scale

is therefore possible to estimate the strength of a salt crystal. One could assume first that under tensile stress all bonds in the crystal planes perpendicular to the applied stress are stretched until they simultaneously break and the material splits into many planes one atom thick – this gives a theoretical strength much greater than in reality. Alternatively, one could assume that only those bonds which are to be broken are stretched – this gives a theoretical strength much less than reality. In practice the true fracture mechanism for these materials turns out to be more involved and more interesting.

A body under tension stores energy – strain energy. The amount of strain energy stored by a brittle material under tension is given by the area under the appropriate stress–strain graph. This strain energy is not uniformly distributed throughout the body but is concentrated around holes, corners and cracks. Inglis (1913) proposed that the stress concentration factor, $K$, around a hole, crack or corner could be calculated according to the formula:

$$K = \left\{ 1 + 2\sqrt{\frac{L}{R}} \right\} \tag{10.1}$$

$L$ = half the length of the crack

$R$ = radius of crack tip or hole

$K$ = stress concentration factor $\left[ \dfrac{\text{local stress}}{\text{mean stress in body}} \right]$

Thus: for a round hole, $K = 3$

for a 2 μm long crack with tip radius equal to half the inter atomic distance

$$(R = 10^{-10} \text{ m}), K = 201$$

Griffith (1921) proposed that for a crack in the surface of a body to propagate the following criteria must be satisfied.

(1) The strain energy that would be released must be greater than the surface energy created.

(2) There must be a crack propagation mechanism available.

Griffith also pointed out that for a given mean stress applied to a body there should be a critical minimum crack length for which the stress concentration at the tip will just be sufficient to cause the crack to propagate. Under the action of this mean stress a crack initially longer than the critical crack length for that stress will grow longer and, since $K$ increases as $L$ increases, crack growth increases until the body is broken. As the crack grows, provided the mean stress remains constant, there is strain energy excess to that required to propagate that crack (since $K$ is increasing). This excess strain energy is dissipated at the velocity of sound in the material to concentrate at the tips of other cracks, causing them to propagate. The rate of crack propagation is lower than the velocity of sound in the material and so other cracks begin to propagate before the first crack brings about failure. Thus, in brittle materials multiple fracture is common. If cracks in the surface of brittle materials can be avoided then the material strength would be near to the theoretical value. This can be demonstrated by heating a glass rod until it softens and then drawing it out to create a new surface. As soon as the rod is cooled it can withstand surprisingly high tensile stress, as demonstrated by bending the rod. Once the new surface is handled or even exposed to the normal environment for a short period, its tensile strength diminishes due to the formation of microscopic cracks in the surface. It has been shown that the surfaces of all materials have cracks in them.

Gilvary (1961) proposed the concept of volume, facial and edge flaws (cracks) in order to calculate the size distribution of breakage products. Assuming that all flaws were randomly distributed and independent of each other and that the initial stress system is removed once the first flaws begin to propagate, Gilvary showed that the product size distributions common to comminuted materials could be predicted. For example, if edge flaws dominate, then the common Rosin–Rammler distribution results.

Evans *et al.* (1961) showed that for a disc acted upon by opposing diametrical loads, there is a uniform tensile stress acting at 90° to the diameter. Under sufficiently high compressive loads, therefore, the resulting tensile stress could exceed the cohesive strength of the material and the disc would split across the diameter. Evans extended the analysis to three-dimensional particles to show that even when particles are stressed compressively, the stress pattern set up by virtue of the shape of the particle may cause it to fail in tension, whether cracks exist or not.

Cracks are less important for 'tough' materials (e.g. rubber, plastics and metals) since excess strain energy is used in deformation of the material rather than crack propagation. Thus in ductile metals, for example the stress concentration at the top of a crack will cause deformation of the material around the crack tip, resulting in a larger tip radius and lower stress concentration.

The observation that small particles are more difficult to break than large particles can be explained using the concept of failure by crack propagation. Firstly, the length of a crack is limited by the size of the particle and so one would expect lower maximum stress concentration factors to be achieved in small particles. Lower stress concentrations mean that higher mean stresses have to be applied to the particles to cause failure. Secondly, the Inglis equation (Equation (10.1)) overpredicts $K$ in the case of small particles since in these particles there is less room for the stress distribution patterns to develop. This effectively limits the maximum stress concentration possible and means that a higher mean stress is necessary to cause crack propagation. Kendal (1978) showed that as particle size decreases, the fracture strength increases until a critical size is reached when crack propagation becomes impossible. Kendal offered a way of predicting this critical particle size.

## 10.3   MODEL PREDICTING ENERGY REQUIREMENT AND PRODUCT SIZE DISTRIBUTION

### 10.3.1   Energy Requirement

There are three well-known postulates predicting energy requirements for particle size reduction. We will cover them in the chronological order in which they were proposed. Rittinger (1867) proposed that the energy required for particle size reduction was directly proportional to the area of new surface created. Thus, if the initial and final particle sizes are $x_1$ and $x_2$ respectively, then assuming a volume shape factor $k_v$ independent of size,

$$\text{volume of initial particle} = k_v x_1^3$$

$$\text{volume of final particle} = k_v x_2^3$$

and each particle of size $x_1$ will give rise to $x_1^3/x_2^3$ particles of size $x_2$.

If the surface shape factor $k_s$ is also independent of size, then for each original particle, the new surface created upon reduction is given by the expression

$$\left(\frac{x_1^3}{x_2^3}\right) k_s x_2^2 - k_s x_1^2 \tag{10.2}$$

which simplifies to

$$k_s x_1^3 \left[\frac{1}{x_2} - \frac{1}{x_1}\right] \tag{10.3}$$

Therefore,

new surface created per unit mass of original particles

$$= k_s x_1^3 \left[\frac{1}{x_2} - \frac{1}{x_1}\right] \times [\text{number of original particles per unit mass}]$$

$$= k_s x_1^3 \left[\frac{1}{x_2} - \frac{1}{x_1}\right] \times \left[\frac{1}{k_v x_1^3 \rho_p}\right] \quad (\text{where } \rho_p \text{ is the particle density})$$

$$= \frac{k_s}{k_v} \frac{1}{\rho_p} \left[\frac{1}{x_2} - \frac{1}{x_1}\right]$$

Hence assuming shape factors and density are constant, Rittinger's postulate may be expressed as

$$\text{breakage energy per unit mass of feed, } E = C_R \left[ \frac{1}{x_2} - \frac{1}{x_1} \right] \tag{10.4}$$

where $C_R$ is a constant. If this is the integral form, then in differential form, Rittinger's postulate becomes

$$\frac{dE}{dx} = -C_R \frac{1}{x^2} \tag{10.5}$$

However, since in practice the energy requirement is usually 200–300 times that required for creation of new surface, it is unlikely that energy requirement and surface created are related.

On the basis of stress analysis theory for plastic deformation, Kick (1885) proposed that the energy required in any comminution process was directly proportional to the ratio of the volume of the feed particle to the product particle. Taking this assumption as our starting point, we see that:

volume ratio, $x_1^3/x_2^3$, determines the energy requirement

(assuming shape factor is constant).

Therefore, size ratio, $x_1/x_2$ fixes the volume ratio, $x_1^3/x_2^3$, which determines the energy requirement. And so, if $\Delta x_1$ is the change in particle size,

$$\frac{x_2}{x_1} = \frac{x_1 - \Delta x_1}{x_1} = 1 - \frac{\Delta x_1}{x_1}$$

which fixes volume ratio, $x_1^3/x_2^3$, and determines the energy requirement.

And so, $\Delta x_1/x_1$ determines the energy requirement for particle size reduction from $x_1$ to $x_1 - \Delta x_1$. Or

$$\Delta E = C_K \left( \frac{\Delta x}{x} \right)$$

As $\Delta x_1 \rightarrow 0$, we have

$$\frac{dE}{dx} = C_K \frac{1}{x} \tag{10.6}$$

This is Kick's law in differential form ($C_K$ is the Kick's law constant). Integrating, we have

$$E = C_K \ln \left( \frac{x_1}{x_2} \right) \tag{10.7}$$

This proposal is unrealistic in most cases since it predicts that the same energy is required to reduce 10 μm particles to 1 μm particles as is required to reduce 1 m boulders to 10 cm blocks. This is clearly not true and Kick's law gives ridiculously low values if data gathered for large product sizes are extrapolated to predict energy requirements for small product sizes.

Bond (1952) suggested a more useful formula, presented in its basic form in Equation (10.8a).

$$E = C_B \left[ \frac{1}{\sqrt{x_2}} - \frac{1}{\sqrt{x_1}} \right]$$ (10.8a)

However, Bond's law is usually presented in the form shown in Equation (10.8b). The law is based on data which Bond obtained from industrial and laboratory scale processes involving many materials.

$$E_B = W_I \left\{ \frac{10}{\sqrt{X_2}} - \frac{10}{\sqrt{X_1}} \right\}$$ (10.8b)

where, $E_B$ = energy required to reduce the top particle size of the material from $x_1$ to $x_2$.

Since top size is difficult to define, in practice $X_1$ to $X_2$ are taken to be the sieve size in microns through which 80% of the material, in the feed and product respectively, will pass. Bond attached particular significance to the 80% passing size.

$W_I$ = the Bond work index. Inspection of Equation (10.8b) reveals that $W_I$ is defined as the energy required to reduce the size of unit mass of material from infinity to 100 μm in size. Although the work index is defined in this way, it is actually determined through laboratory scale experiment and assumed to be independent of final product size.

Both $E_B$ and $W_I$ have the dimensions of energy per unit mass and commonly expressed in the units kilowatt-hour per short ton (2000 lb) (1 kWh/short ton ≈ 4000 J/kg). The Bond work index, $W_I$, must be determined empirically. Some common examples are: bauxite, 9.45 kWh/short ton; coke from coal, 20.7 kWh/short ton; gypsum rock, 8.16 kWh/short ton.

Bond's formula gives a fairly reliable first approximation to the energy requirement provided the product top size is not less than 100 μm. In differential form Bond's formula becomes:

$$\frac{dE}{dx} = C_B \frac{1}{x^{3/2}}$$ (10.9)

Attempts have been made (e.g. Holmes, 1957; Hukki, 1961) to find the general formula for which the proposals of Rittinger, Kick and Bond are special cases. It can be seen from the results of the above analysis that these three proposals can be considered as being the integrals of the same differential equation:

$$\frac{\mathrm{d}E}{\mathrm{d}x} = -C\frac{1}{x^N}$$                                (10.10)

with

$N = 2$        $C = C_R$        for Rittinger

$N = 1$        $C = C_K$        for Kick

$N = 1.5$        $C = C_B$        for Bond

It has been suggested that three approaches to prediction of energy requirements mentioned above are each more applicable in certain areas of product
size. It is common practice to assume that Kick's proposal is applicable for
large particle size (coarse crushing and crushing), Rittinger's for very small
particle size (ultra-fine grinding) and the Bond formula being suitable for
intermediate particle size – the most common range for many industrial
grinding processes. This is shown in Figure 10.2, in which specific energy
requirement is plotted against particle size on logarithmic scales. For Rittinger's postulate, $E \propto 1/x$ and so $\ln E \propto -1\ln(x)$ and hence the slope is $-1$. For
Bond's formula, $E \propto 1/x^{0.5}$ and so $\ln E \propto -0.5\ln(x)$ and hence the slope is
$-0.5$. For Kick's law, the specific energy requirement is dependent on the
reduction ratio $x_1/x_2$ irrespective of the actual particle size; hence the slope is
zero.

  In practice, however, it is generally advisable to rely on the past experience
of equipment manufacturers and on tests in order to predict energy requirements for the milling of a particular material.

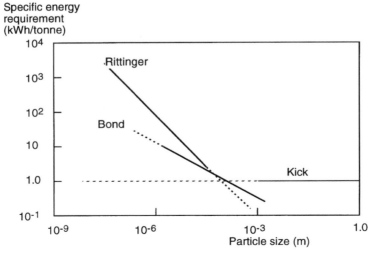

**Figure 10.2**  Specific energy requirement for breakage: relationship to laws of Rittinger,
Bond and Kick

## 10.3.2   Prediction of the Product Size Distribution

It is common practice to model the breakage process in comminution equipment on the basis of two functions, the specific rate of breakage and the breakage distribution function. The specific rate of breakage $S_j$ is the probability of a particle of size $j$ being broken in unit time (in practice, 'unit time' may mean a certain number of mill revolutions, for example). The breakage distribution function $b(i, j)$ describes the size distribution of the product from the breakage of a given size of particle. For example, $b(i, j)$ is the fraction of breakage product from size interval $j$ which falls into size interval $i$. Figure 10.3 helps demonstrate the meaning of $S_j$ and $b(i, j)$ when dealing with 10 kg of monosized particles in size interval 1. If $S_1 = 0.6$ we would expect 4 kg of material to remain in size interval 1 after unit time. The size distribution of the breakage product would be described by the set of $b(i, j)$ values. Thus, for example, if $b(4, 1) = 0.25$ we would expect to find 25% by mass from size interval 1 to fall into size interval 4. The breakage distribution function may also be expressed in cumulative form as $B(i, j)$, the fraction of the breakage product from size interval $j$ which falls into size intervals $j$ to $n$, where $n$ is the total number of size intervals. ($B(i, j)$ is thus a cumulative undersize distribution.)

Thus, remembering that $S$ is a *rate* of breakage, we have Equation (10.11), which expresses the rate of change of the mass of particles in size interval $i$ with time:

$$\frac{\mathrm{d}m_i}{\mathrm{d}t} = \sum_{j=1}^{j=i-1} \{b(i, j)S_j m_j\} - S_i m_i \qquad (10.11)$$

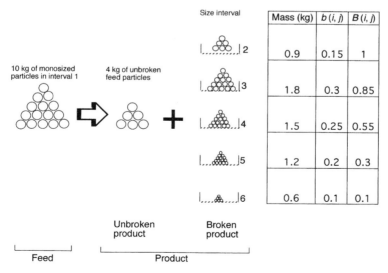

| Size interval | Mass (kg) | $b\,(i, j)$ | $B\,(i, j)$ |
|---|---|---|---|
| 2 | 0.9 | 0.15 | 1 |
| 3 | 1.8 | 0.3 | 0.85 |
| 4 | 1.5 | 0.25 | 0.55 |
| 5 | 1.2 | 0.2 | 0.3 |
| 6 | 0.6 | 0.1 | 0.1 |

10 kg of monosized particles in interval 1

4 kg of unbroken feed particles

Unbroken product        Broken product

Feed            Product

**Figure 10.3**   The meanings of specific rate of breakage and breakage distribution function

where

$$\sum_{j=1}^{j=i-1} \{b(i, j)S_j m_j\} = \text{mass broken into interval } i \text{ from all intervals of } j > i$$

$$S_i m_i = \text{mass broken out of interval } i$$

Since $m_i = y_i M$ and $m_j = y_j M$, where $M$ is the total mass of feed material and $y_i$ is the mass fraction in size interval $i$, then we can write a similar expression for the rate of change of mass fraction of material in size interval, $i$ with time (Equation (10.12)).

$$\frac{dy_i}{dt} = \sum_{j=1}^{j=i-1} \{b(i, j)S_j y_j\} - S_i y_i \tag{10.12}$$

Thus, with a set of $S$ and $b$ values for a given feed material, the product size distribution after a given time in a mill may be determined. In practice, both $S$ and $b$ are dependent on particle size, material and machine. From the earlier discussion on particle fracture mechanisms it would be expected that the specific rate of breakage should decrease with decreasing particle size, and this is found to be the case. The aim of this approach is to be able to use values of $S$ and $b$ determined from small-scale tests to predict product size distributions on a large scale. This method is found to give fairly reliable predictions.

## 10.4   TYPES OF COMMINUTION EQUIPMENT

### 10.4.1   Factors Affecting Choice of Machine

The choice of machine selected for a particular grinding operation will depend on the following variables:

- stressing mechanism;
- size of feed and product;
- material properties;
- carrier medium;
- mode of operation;
- capacity;
- combination with other unit operations.

## 10.4.2   Stressing Mechanisms

It is possible to identify three stresses mechanisms responsible for particle size reduction in mills.

(1) Stress applied between two surfaces (either surface–particle or particle–particle) at low velocity, 0.01–10 m/s. Crushing plus attrition, Figure 10.4.

(2) Stress applied at a single solid surface (surface–particle or particle–particle) at high velocity, 10–200 m/s. Impact fracture plus attrition, Figure 10.5.

(3) Stress applied by carrier medium—usually in wet grinding to bring about disagglomeration.

An initial classification of comminution equipment can be made according to the stressing mechanisms employed, as follows.

*Machines using mainly mechanism 1, crushing*

The *jaw crusher* behaves like a pair of giant nutcrackers (Figure 10.6). One jaw is fixed and the other, which is hinged at its upper end, is moved towards and away from the fixed jaw by means of toggles driven by an eccentric. The lumps of material are crushed between the jaws and leave the crusher when they are able to pass through a grid at the bottom.

    The *gyratory crusher*, shown in Figure 10.7, has a fixed jaw in the form of a truncated cone. The other jaw is a cone which rotates inside the fixed jaw on an eccentric mounting. Material is discharged when it is small enough to pass through the gap between the jaws.

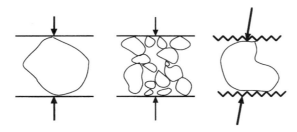

**Figure 10.4**   Stresses applied between two surfaces

**Figure 10.5**   Stresses applied at a single solids surface

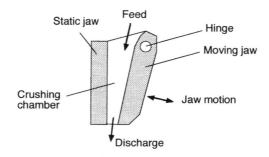

**Figure 10.6**   Schematic diagram of a jaw crusher

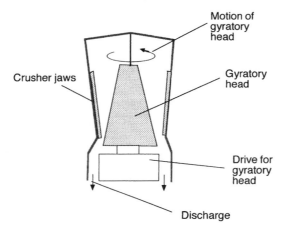

**Figure 10.7**   Schematic diagram of gyratory crusher

In the *crushing roll* machine two cylindrical rolls rotate in opposite directions, horizontally and side by side with an adjustable gap between them (Figure 10.8). As the rolls rotate the drag in material which is choke-fed by gravity so that particle fracture occurs as the material passes through the gap

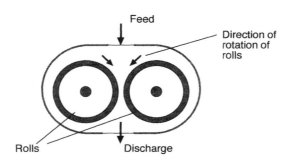

**Figure 10.8**   Schematic diagram of crushing oil rolls

between the rolls. The rolls may be ribbed to give improved purchase between material and rolls.

In the *horizontal table mill*, shown in Figure 10.9, the feed material falls on to the centre of a circular rotating table and is thrown out by centrifugal force. In moving outwards the material passes under a roller and is crushed.

## Machines using mainly mechanism 2, high velocity impact

The *hammer mill*, shown in Figure 10.10, consists of a rotating shaft to which are attached fixed or pivoted hammers. This device rotates inside a cylinder. The particles are fed into the cylinder either by gravity or by gas stream. In the gravity-fed version the particles leave the chamber when they are small enough to pass through a grid at the bottom.

A *pin mill* consists of two parallel circular discs each carrying a set of projecting pins (Figure 10.11). One disc is fixed and the other rotates at high speed so that its pins pass close to those on the fixed disc. Particles are carried in air into the centre and as they move radially outwards are fractured by impact or by attrition.

The *fluid energy mill* relies on the turbulence created in high velocity jets of air or steam in order to produce conditions for inter-particle collisions which bring about particle fracture. A common form of fluid energy mill developed since the Second World War is the loop or oval jet mill shown in Figure 10.12. Material is conveyed from the grinding area near the jets at the base of the loop to the classifier and exit situated at the top of the loop. These mills have a very high specific energy consumption and are subject to extreme wear when handling abrasive materials. These problems have been overcome to a certain extent in the recently developed fluidized bed jet mill in which the bed is used to absorb the energy from the high-speed particles ejected from the grinding zone.

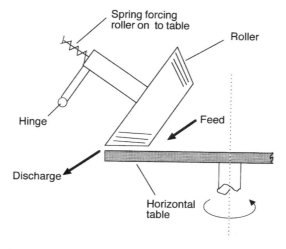

**Figure 10.9** Schematic diagram of horizontal table mill

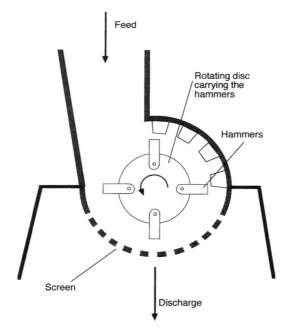

**Figure 10.10**   Schematic diagram of a hammer mill

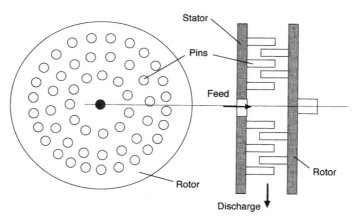

**Figure 10.11**   Schematic diagram of a pin mill

*Machines using a combination of mechanisms 1 and 2, crushing and impact with attrition*

The *Sand Mill*, shown in Figure 10.13, is a vertical cylinder containing a stirred bed of sand, glass beads or shot. The feed, in the form of a slurry, is pumped into the bottom of the bed and the product passes out at the top through a screen which retains the bed material.

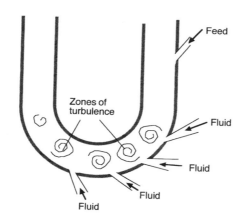

**Figure 10.12**    Schematic diagram of a fluid energy mill

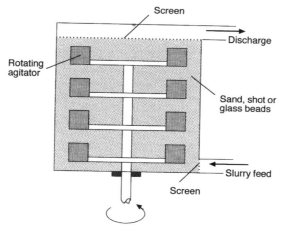

**Figure 10.13**    Schematic diagram of a sand mill

In the *Colloid Mill*, the feed in the form of a slurry passes through the gap between a male, ribbed cone rotating at high speed and a female static cone (Figure 10.14).

The *ball mill*, shown in Figure 10.15, is a rotating cylindrical or cylindrical–conical shell about half filled with balls of steel or ceramic. The speed of rotation of the cylinder is such that the balls are caused to tumble over one another without causing cascading. This speed is usually less than 80% of the critical speed which would just cause the charge of balls and feed material to be centrifuged. In continuous milling the carrier medium is air, which may be heated to avoid moisture which tends to cause clogging. Ball mills may also be used for wet grinding with water being used as the carrier medium. The size of balls is chosen to suit the desired product size. The conical section of the

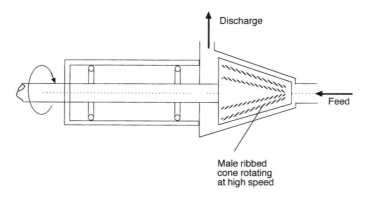

**Figure 10.14**   Schematic diagram of a colloid mill

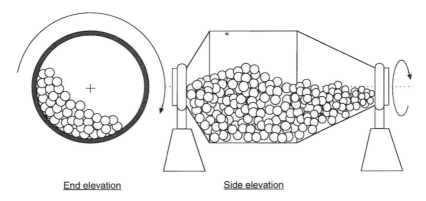

End elevation                        Side elevation

**Figure 10.15**   Schematiac diagram of a ball mill

mill shown in Figure 10.15 causes the smaller balls to move towards the discharge end and accomplish the fine grinding. Tube mills are very long ball mills which are often compartmented by diaphragms, with balls graded along the length from large at the feed end to small at the discharge end.

### 10.4.3   Particle Size

Although it is technically interesting to classify mills according to the stressing mechanisms it is the size of the feed and the product size distribution required which in most cases determine the choice of a suitable mill. Generally the terminology shown in Table 10.1 is used.

Table 10.2 indicates how the product size determines the type of mill to be used.

**Table 10.1**  Terminology used in comminution

| Size range of product | Term used |
|---|---|
| 1–0.1 m | Coarse crushing |
| 0.1 m | Crushing |
| 1 cm | Fine crushing, coarse grinding |
| 1 mm | Intermediate grinding, milling |
| 100 μm | Fine grinding |
| 10 μm | Ultrafine grinding |

**Table 10.2**  Categorizing comminution equipment according to product size

| Down to 3 mm | 3 mm–50 μm | < 50 μm |
|---|---|---|
| Crushers | Ball mills | Ball mills |
| Table mills | Rod mills | Vibration mills |
| Edge runner mills | Pin mills | Sand mills |
| | Tube mills | Perl mills |
| | Vibration mills | Colloid mills |
| | | Fluid energy mills |

## 10.4.4  Material Properties

Material properties affect the selection of mill type, but to a lesser extent than feed and product particle size. The following material properties may need to be considered when selecting a mill:

- *Hardness*. Hardness is usually measured on the Mohs' scale of hardness where graphite is ranked 1 and diamond is ranked 10. The property hardness is a measure of the resistance to abrasion.

- *Abrasiveness*. This is linked closely to hardness and is considered by some to be the most important factor in selection of commercial mills. Very abrasive materials must generally be ground in mills operating at low speeds to reduce wear of machine parts in contact with the material (e.g. ball mills).

- *Toughness*. This is the property whereby the material resists the propagation of cracks. In tough materials excess strain energy brings about plastic deformation rather than propagation of new cracks. Brittleness is the opposite of toughness. Tough materials present problems in grinding, although in some cases it is possible to reduce the temperature of the material, thereby reducing the propensity to plastic flow and rendering the material more brittle.

- *Cohesivity/adhesivity.* The properties whereby particles of material stick together and to other surfaces. Cohesivity and adhesivity are related to moisture content and particle size. Decrease of particle size or increasing moisture content increases the cohesivity and adhesivity of the material. Problems caused by cohesivity/adhesivity due to particle size may be overcome by wet grinding.

- *Fibrous nature.* Materials of a fibrous nature are a special case and must be comminuted in shredders or cutters which are based on the hammer mill design.

- *Low melting point.* The heat generated in a mill may be sufficient to cause melting of such materials causing problems of increased toughness and increased cohesivity and adhesivity. In some cases, the problem may be overcome by using cold air as the carrier medium.

- *Other special properties.* Materials which are thermally sensitive and have a tendency to spontaneous combustion or high inflammability must be ground using an inert carrier medium (e.g. nitrogen). Toxic or radioactive materials must be ground using a carrier medium operating on a closed circuit.

## 10.4.5   Carrier Medium

The carrier medium may be a gas or a liquid. Although the most common gas used is air, inert gases may be used in some cases as indicated above. The most common liquid used in wet grinding is water although oils are sometimes used. The carrier medium not only serves to transport the material through the mill but, in general, transmits forces to the particles, influences friction and hence abrasion, affects crack formation and cohesivity/adhesivity. The carrier medium can also influence the electrostatic charging and the flammability of the material.

## 10.4.6   Mode of Operation

Mills operate in either batch or continuous mode. Choice between modes will be based on throughput, the process and economics. The capacity of batch mills varies from a few grammes on the laboratory scale to a few tonnes on a commercial scale. The throughput of continuous milling systems may vary from several hundred grammes per hour at laboratory scale to several thousand tonnes per hour at industrial scale.

## 10.4.7   Combination with other Operations

Some mills have a dual purpose and thus may bring about drying, mixing or classification of the material in addition to its size reduction.

## 10.4.8   Types of Milling Circuit

Milling circuits are either 'open circuit' or 'closed circuit'. In open circuit milling (Figure 10.16) the material passes only once through the mill, and so the only controllable variable is the residence time of the material in the mill. Thus the product size and distribution may be controlled over a certain range by varying the material residence time (thoughput), i.e. feed rate governs product size and so the system is inflexible.

In closed circuit milling (Figure 10.17) the material leaving the mill is subjected to some form of classification (separation according to particle size) with the oversize being returned to the mill with the feed material. Such a system is far more flexible since both product mean size and size distribution may be controlled.

Figures 10.18 and 10.19 show the equipment necessary for feeding material into the mill, removing material from the mill, classifying, recycling oversize material and removing product in the case of dry and wet closed milling circuits respectively.

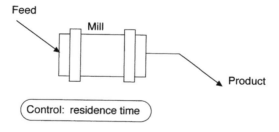

**Figure 10.16**   Open circuit milling

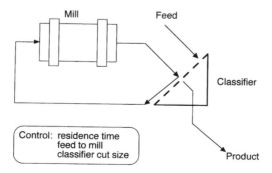

**Figure 10.17**   Closed circuit milling

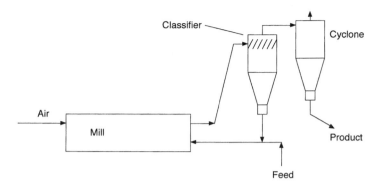

**Figure 10.18**   Dry milling: closed circuit operation

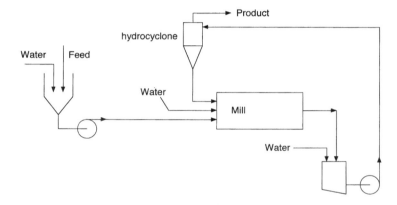

**Figure 10.19**   Wet mill: closed circuit operation

## 10.5   WORKED EXAMPLES

### WORKED EXAMPLE 10.1

A material consisting originally of 25 mm particles is crushed to an average size of 7 mm and requires 20 kJ/kg for this size reduction. Determine the energy required to crush the material from 25 mm to 3.5 mm assuming (a) Rittinger's law, (b) Kick's law and (c) Bond's law.

### Solution

(a) Applying Rittinger's law as expressed by Equation (10.4):

$$20 = C_R \left[ \frac{1}{7} - \frac{1}{25} \right]$$

hence $C_R = 194.4$ and so with $x_2 = 3.5$ mm,

$$E = 194.4 \left[ \frac{1}{3.5} - \frac{1}{25} \right]$$

hence, $E = 47.8$ kJ/kg

(b) Applying Kick's law as expressed by Equation (10.7):

$$20 = -C_K \ln \left[ \frac{7}{25} \right]$$

hence, $C_K = 15.7$ and so with $x_2 = 3.5$ mm,

$$E = -15.7 \ln \left[ \frac{3.5}{25} \right]$$

hence, $E = 30.9$ kJ/kg

(c) Applying Bond's law as expressed by Equation (10.8a):

$$20 = C_B \left[ \frac{1}{\sqrt{7}} - \frac{1}{\sqrt{25}} \right]$$

hence, $C_B = 112.4$ and so with $x_2 = 3.5$ mm,

$$E = 112.4 \left[ \frac{1}{\sqrt{3.5}} - \frac{1}{25} \right]$$

hence, $E = 37.6$ kJ/kg

## WORKED EXAMPLE 10.2

Values of breakage distribution function $b(i, j)$ and specific rates of breakage $S_j$ for a particular material in a ball mill are shown in Table 10W2.1. To test the validity of these values, a sample of the material with the size distribution indicated in Table W2.2 is to be ground in a ball mill. Use the information in these tables to

**Table 10W2.1** Specific rates of breakage and breakage distribution function for the ball mill

| Size interval (µm) | 212–150 | 150–106 | 106–75 | 75–53 | 53–37 | 37–0 |
|---|---|---|---|---|---|---|
| Interval No. | 1 | 2 | 3 | 4 | 5 | 6 |
| $S_j$ | 0.7 | 0.6 | 0.5 | 0.35 | 0.3 | 0 |
| $b(1, j)$ | 0 | 0 | 0 | 0 | 0 | 0 |
| $b(2, j)$ | 0.32 | 0 | 0 | 0 | 0 | 0 |
| $b(3, j)$ | 0.3 | 0.4 | 0 | 0 | 0 | 0 |
| $b(4, j)$ | 0.14 | 0.2 | 0.5 | 0 | 0 | 0 |
| $b(5, j)$ | 0.12 | 0.2 | 0.25 | 0.6 | 0 | 0 |
| $b(6, j)$ | 0.12 | 0.2 | 0.25 | 0.4 | 1.0 | 0 |

**Table 10W2.2**   Feed size distribution

| Interval No. ($j$) | 1 | 2 | 3 | 4 | 5 | 6 |
|---|---|---|---|---|---|---|
| Fraction | 0.2 | 0.4 | 0.3 | 0.06 | 0.04 | 0 |

predict the size distribution of the product after one minute in the mill. Note: $S_j$ values in Table W2.1 are based on one minute grinding time.

### Solution

Applying Equation (10.12):

*Change of fraction in interval 1*

$$\frac{dy_1}{dt} = 0 - S_1 y_1 = 0 - 0.7 \times 0.2$$

$$= -0.14$$

Hence, new $y_1 = 0.2 - 0.14 = 0.06$

*Change of fraction in interval 2*

$$\frac{dy_2}{dt} = b(2, 1)S_1 y_1 - S_2 y_2$$

$$= (0.32 \times 0.7 \times 0.2) - (0.6 \times 0.4)$$

$$= -0.1952$$

Hence new $y_2 = 0.4 - 0.1952 = 0.2048$

*Change in fraction in interval 3*

$$\frac{dy_3}{dt} = [b(3, 1)S_1 y_1 + b(3, 2)S_2 y_2] - S_3 y_3$$

$$= [(0.3 \times 0.7 \times 0.2) + (0.4 \times 0.6 \times 0.4)] - (0.5 \times 0.3)$$

$$= -0.012$$

Hence, new $y_3 = 0.3 - 0.012 = 0.288$

Similarly for intervals 4, 5 and 6:

$$\text{new } y_4 = 0.1816$$

$$\text{new } y_5 = 0.1429$$

$$\text{new } y_6 = 0.1227$$

Checking:
Sum of predicted product interval mass fractions $= y_1 + y_2 + y_3 + y_4 + y_5 + y_6 = 1.000$
Hence product size distribution:

| Interval No. ($j$) | 1 | 2 | 3 | 4 | 5 | 6 |
|---|---|---|---|---|---|---|
| Fraction | 0.06 | 0.2048 | 0.288 | 0.1816 | 0.1429 | 0.1227 |

## EXERCISES

**10.1**
(a) Rittinger's energy law postulated that the energy expended in crushing is proportional to the area of new surface created. Derive an expression relating the specific energy consumption in reducing the size of particles from $x_1$ to $x_2$ according to this law.

(b) Table 10E1.1 below gives values of specific rates of breakage and breakage distribution functions for the grinding of limestone in a hammer mill. Given that values of specific rates of breakage are based on 30 s in the mill at a particular speed, determine the size distribution of the product resulting from the feed described in Table 10E1.2 after 30 s in the mill at this speed.

(Answer: 0.12, 0.322, 0.314, 0.244)

**10.2** Table 10E2.1 gives information gathered from tests on the size reduction of coal in a ball mill. Assuming that the values of specific rates of breakage, $S_j$, are based on 25 revolutions of the mill at a particular speed, predict the product size

**Table 10E1.1**  Specific rates of breakage and breakage distribution function for the hammer mill

| Interval (µm) | 106–75 | 75–53 | 53–37 | 37–0 |
|---|---|---|---|---|
| Interval No. $j$ | 1 | 2 | 3 | 4 |
| $S_j$ | 0.6 | 0.5 | 0.45 | 0 |
| $b(1, j)$ | 0 | 0 | 0 | 0 |
| $b(2, j)$ | 0.4 | 0 | 0 | 0 |
| $b(3, j)$ | 0.3 | 0.6 | 0 | 0 |
| $b(4, j)$ | 0.3 | 0.4 | 1.0 | 0 |

**Table 10E1.2**  Feed size distribution

| Interval | 1 | 2 | 3 | 4 |
|---|---|---|---|---|
| Fraction | 0.3 | 0.5 | 0.2 | 0 |

**Table 10E2.1**    Results of ball mill tests on coal

| Interval (μm) | 300–212 | 212–150 | 150–106 | 106–75 | 75–53 | 53–37 | 37–0 |
|---|---|---|---|---|---|---|---|
| Interval no. $j$ | 1 | 2 | 3 | 4 | 5 | 6 | 7 |
| $S_j$ | 0.5 | 0.45 | 0.42 | 0.4 | 0.38 | 0.25 | 0 |
| $b(1, j)$ | 0 | 0 | 0 | 0 | 0 | 0 | 0 |
| $b(2, j)$ | 0.25 | 0 | 0 | 0 | 0 | 0 | 0 |
| $b(3, j)$ | 0.24 | 0.29 | 0 | 0 | 0 | 0 | 0 |
| $b(4, j)$ | 0.19 | 0.27 | 0.33 | 0 | 0 | 0 | 0 |
| $b(5, j)$ | 0.12 | 0.2 | 0.3 | 0.45 | 0 | 0 | 0 |
| $b(6, j)$ | 0.1 | 0.16 | 0.25 | 0.3 | 0.6 | 0 | 0 |
| $b(7, j)$ | 0.1 | 0.08 | 0.12 | 0.25 | 0.4 | 1.0 | 0 |

distribution resulting from the feed material, details of which are given in Table 10E2.2 after 25 revolutions in the mill at that speed.

(Answer: 0.125, 0.2787, 0.2047, 0.1661, 0.0987, 0.0779, 0.04878)

**10.3** Table 10E3.1 gives information on the size reduction of a sand-like material in a ball mill. Given that the values of specific rates of breakage $S_j$ are based on 5 revolutions of the mill, determine the size distribution of the feed materials shown in Table 10E3.2 after 5 revolutions of the mill.

(Answer: 0.0875, 0.2369, 0.2596, 0.2115, 0.2045)

**10.4** Comminution processes are generally less than 1% efficient. Where does all the wasted energy go?

**Table 10E2.2**    Feed size distribution

| Interval | 1 | 2 | 3 | 4 | 5 | 6 | 7 |
|---|---|---|---|---|---|---|---|
| Fraction | 0.25 | 0.45 | 0.2 | 0.1 | 0 | 0 | 0 |

**Table 10E3.1**    Results of ball mill tests

| Interval (μm) | 150–106 | 106–75 | 75–53 | 53–37 | 37–0 |
|---|---|---|---|---|---|
| Interval No. $(j)$ | 1 | 2 | 3 | 4 | 5 |
| $S_j$ | 0.65 | 0.55 | 0.4 | 0.35 | 0 |
| $b(1, j)$ | 0 | 0 | 0 | 0 | 0 |
| $b(2, j)$ | 0.35 | 0 | 0 | 0 | 0 |
| $b(3, j)$ | 0.25 | 0.45 | 0 | 0 | 0 |
| $b(4, j)$ | 0.2 | 0.3 | 0.6 | 0 | 0 |
| $b(5, j)$ | 0.2 | 0.25 | 0.4 | 1.0 | 0 |

**Table 10E3.2** Feed size distribution

| Interval | 1 | 2 | 3 | 4 | 5 |
|----------|------|-----|-----|-----|------|
| Fraction | 0.25 | 0.4 | 0.2 | 0.1 | 0.05 |

# 11

# Size Enlargement

## 11.1 INTRODUCTION

Size enlargement is the process by which smaller particles are put together to form larger masses in which the original particles can still be identified. Size enlargement is one of the single most important process steps involving particulate solids in the process industries. Size enlargement is mainly associated with the pharmaceutical, agricultural and food industries, but also plays an important role in other industries including minerals, metallurgical and ceramics.

There are many reasons why we may wish to increase the mean size of a product or intermediate. These include reduction of dust hazard (explosion hazard or health hazard), to reduce caking and lump formation, to improve flow properties, increase bulk density for storage, creation of non-segregating mixtures of ingredients of differing original size, to provide a defined metered quantity of active ingredient (e.g. pharmaceutical drug formulations), control of surface to volume ratio (e.g. in catalyst supports).

Methods by which size enlargement is brought about include granulation, compaction (e.g. tabletting), extrusion, sintering, spray drying and prilling. Agglomeration is the formation of agglomerates or aggregates by sticking together of smaller particles and granulation is agglomeration by agitation methods. Since it is not possible in the confines of a single chapter to cover all size enlargement methods adequately, the focus of this chapter will be on granulation, which will serve as an example.

In this chapter the different types of interparticle forces and their relative importance as a function of particle size are summarized. Liquid bridge forces, which are specifically important to granulation processes, are covered in more detail. The rate processes important to granulation (wetting, growth, consolidation and attrition) are reviewed and population balance included in order to develop a simple model for simulation for the granulation process. To conclude, a brief overview of industrial granulation equipment is also given.

## 11.2   INTERPARTICLE FORCES

### 11.2.1   Van der Waals Forces

There exist between all solids molecularly based attractive forces collectively known as van der Waals forces. The energy of these forces is of the order of 0.1 electron-volt and decreases with the sixth power of the distance between molecules. The range of van der Waals forces is large compared with that of chemical bonds. The attractive force, $F_{vw}$, between a sphere and a plane surface as a result of van der Waals forces was derived by Hamaker (1937) and is usually presented in the form

$$F_{vw} = \frac{K_H R}{6y^2} \tag{11.1}$$

where $K_H$ is the Hamaker constant, $R$ is the radius of the sphere and $y$ is the gap between the sphere and the plane.

### 11.2.2   Forces due to Adsorbed Liquid Layers

Particles in the presence of a condensable vapour will have a layer of adsorbed vapour on their surface. If these particles are in contact a bonding forces results from the overlapping of the adsorbed layers. The strength of the bond is dependent on the area of contact and the tensile strength of the adsorbed layers. The thickness and strength of the layers increase with increasing partial pressure of the vapour in the surrounding atmosphere. According to Coelho and Harnby (1978) there is a critical partial pressure at which the adsorbed layer bonding gives way to liquid bridge bonding.

### 11.2.3   Forces due to Liquid Bridges

In addition to the interparticle forces resulting from adsorbed liquid layers, as described above, even in very small proportions the presence of liquid on the surface of particles affects the interparticle forces by the smoothing effect it has on surface imperfections (increasing particle–particle contact) and its effect of reducing the interparticle distance. However, these forces are usually negligible in magnitude compared with forces resulting when the proportion of liquid present is sufficient to form interparticle liquid bridges. Newitt and Conway-Jones (1958) identified four types of liquid states depending on the proportion of liquid present between groups of particles. These states are known as pendular, funicular, capillary and droplet. These are shown in Figure 11.1. In the pendular state liquid is held as a point contact in a bridge neck between particles. The liquid bridges are all separate and independent of each other. The strong boundary forces resulting from the surface tension of the liquid draw the particles together. In addition there is capillary pressure resulting from the curve liquid surfaces of the bridge. The capillary pressure is given by

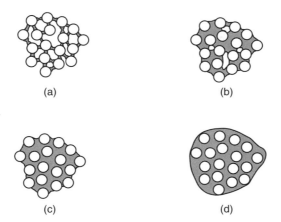

**Figure 11.1**   Liquid bonding between particles: (a) pendular; (b) funicular; (c) capillary; (d) droplet

$$p_c = \gamma \left[ \frac{1}{r_1} - \frac{1}{r_2} \right]$$                                          (11.2)

where $\gamma$ is the liquid surface tension and $r_1$ and $r_2$ are the radii of curvature of the liquid surfaces. If the pressure in the liquid bridge is less than the ambient pressure, the capillary pressure and surface tensions boundary attractive force are additive.

As the proportion of liquid to particles is increased the liquid is free to move and the attractive force between particles decreases (funicular). When there is sufficient liquid to completely fill the interstices between the particles (capillary) the granule strength falls further as there are fewer curved liquid surfaces and fewer boundaries for surface tensions forces to act on. Clearly when the particles are completely dispersed in the liquid (droplet) the strength of the structure is very low.

In the pendular state increasing the amount of liquid present has little effect on the strength of the bond between the particles until the funicular state is achieved. However, increasing the proportion of liquid increases the resistance of the bond to rupture since the particles can be pulled further apart without rupture of bridges. This has practical implication for granulation processes; pendular bridges give rise to strong granules in which the quantity of liquid is not critical but should be less than that required to move into the funicular and capillary regimes.

## 11.2.4   Electrostatic Forces

Electrostatic charging of particles and surfaces occurs as a result of friction caused by interparticle collisions and frequent rubbing of particles against equipment surfaces during processing. The charge is caused by the transfer of electrons between the bodies. The force between two charged spheres is proportional to the product of their charges. Electrostatic forces may be attractive or repulsive, do not require contact between particles and can act over relatively long distances compared to adhesional forces which require contact.

## 11.2.5 Solid Bridges

Granules formed by liquid bridges are usually not the end product in a granulation process. More permanent bonding within the granule is created by solid bridges formed as liquid is removed from the original granule. Solid bridges between particles may take three forms: crystalline bridges, liquid binder bridges and solid binder bridges. If the material of the particles is soluble in the liquid added to create granules, crystalline bridges may be formed when the liquid evaporates. The process of evaporation reduces the proportion of liquid in the granules producing high strength pendular bridges before crystals form. Alternatively the liquid used initially to form the granules may contain a binder or glue which takes effect upon evaporation of solvent.

In some cases a solid binder may be used. This is a finely ground solid which reacts with the liquid present to produce a solid cement to hold the particle together.

## 11.2.6 Comparison and Interaction between Forces

In practice all interparticle forces act simultaneously. The relative importance of the forces varies with changes in particle properties and with changes in the humidity of the surrounding atmosphere. There is considerable interaction between the bonding forces. For example, in aqueous systems adsorbed moisture can considerably increase van der Waals forces. Adsorbed moisture can also reduce interparticle friction and potential for interlocking, making the powder more free-flowing. Electrostatic forces decay rapidly if the humidity of the surrounding air is increased.

A powder which in a dry atmosphere exhibits cohesivity due to electrostatic charging may become more free-flowing as humidity of the atmosphere is increased. If humidity is further increased liquid bridge formation can result in a return to cohesive behaviour. This effect has been reported in powder mixing studies by Coelho and Harnby (1978) and Karra and Fuerstenau (1977).

Figure 11.2, after Rumpf (1962), illustrates the relative magnitude of the different bonds discussed above as a function of particle size. We see that van der Waals forces become important only for particles below 1 μm in size, adsorbed vapour forces are relevant below 80 μm and liquid bridge forces are active below about 500 μm.

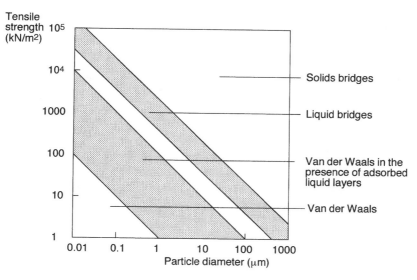

**Figure 11.2**  Theoretical tensile strength of agglomerates with different bonding mechanisms (after Rumpf, 1962)

## 11.3  GRANULATION

### 11.3.1  Introduction

Granulation is particle size enlargement by sticking together smaller particles using agitation methods to distribute liquid binder and impart energy to particles and granules. The motion of the particles and granules results in collisions which produce growth by coalscence and coating. In general the individual component particles are held together by a liquid (glue-like) or solid (cement-like) binder which is sprayed into the agitated particles in the granulator. Alternatively, a solvent may be used to induce dissolution and recrystallization of the material of which the particles are made.

### 11.3.2  Granulation Rate Processes

In granulation the formation of granules or agglomerates is controlled by four rate processes. These are *wetting* of the original particles by the binding liquid, coalescence or *growth* to form granules, *consolidation* of the granule and attrition or *breakage* of the granule. These processes combine to determine the properties of the product granule (size distribution, porosity, strength, dispersibility etc.). The final granule size in a granulation process is controlled by the competing mechanisms of growth, breakage and consolidation.

## Wetting

Wetting is the process by which air within the voids between particles is replaced by liquid. Ennis and Litster (1997) stress the importance influence which the extent and rate of *wetting* have on product quality in a granulation process. For example, poor wetting can result in much material being left ungranulated and requiring recycling. When granulation is used to combine ingredients, account must be taken of the different wetting properties which components of the final granule may have.

Wetting is governed by the surface tension of the liquid and the contact angle it forms with the material of the particles. The rate at which wetting occurs is important in granulation. An impression of this rate is given by the Washburn equation (Equation (11.3)) for the rate of penetration of liquid of viscosity $\mu$ and surface tension $\gamma$ into a bed of powder when gravity is not significant:

$$\frac{dz}{dt} = \frac{R_p \gamma \cos \theta}{4\mu z} \tag{11.3}$$

where $t$ is the time, $z$ is the penetration of the liquid into the powder and $\theta$ is the dynamic contact angle of the liquid with the solid of the powder. $R_p$ is the average pore radius which is related to the packing density and the size distribution of the powder. Thus in granulation, the factors controlling the rate of wetting are adhesive tension ($\gamma \cos \theta$), liquid viscosity, packing density and the size distribution.

In general, improved wetting is desirable. It gives a narrower granule size distribution and improved product quality through better control over the granulation process. In practice, the rate of wetting will significantly influence the extent of wetting of the powder mass, especially where evaporation of binder solvent takes place simultaneously with wetting. The brief analysis above shows us that the rate of wetting is increased by reducing viscosity, increasing surface tension, minimizing contact angle and increasing the size of pores within the powder. Viscosity is determined by the binder concentration and the operating temperature. As concentration changes with solvent evaporation, the binder viscosity will increase. Small particles give small pores and large particles give large pores. Also, a wider particle size distribution will give rise to smaller pores. Large pores ensure a high rate of liquid penetration but give rise to a lower extent of wetting.

## Growth

The mechanisms controlling the rate of growth of granules were identified by Sastry and Fuerstenau (1977) and are summarized in Figure 11.3. The growth mechanisms are

(1) nucleation;

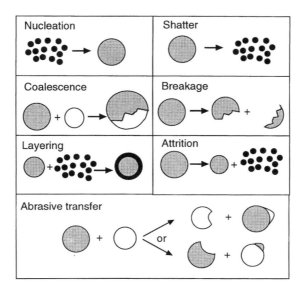

**Figure 11.3** Summary of the mechanisms controlling the growth rate of granules (after Sastry and Fuerstenau, 1977)

(2) coalescence;

(3) layering or coating.

Nucleation is the term used to describe the initial process of combination of primary solid particles into a liquid drop to form primary granules or nuclei. Coalescence is the joining together of these primary granules to form larger granules. Layering is the addition of primary particles to primary or larger granules in layers.

For two colliding primary granules to coalesce their kinetic energy must be dissipated and the strength of the resulting bond must be able to resist the external forces exerted by the agitation of the powder mass in the granulator. Granules which are able to deform readily will absorb the collisional energy and create increased surface area for bonding. As granules grow so do the internal forces trying to pull the granule apart. It is possible to predict a critical maximum size of granule beyond which coalescence is not possible during collision (see below).

Ennis and Litster (1997) suggest a rationale for interpreting observed granule growth regimes in terms of collision physics. Consider collision between two rigid granules (assume low deformability) of density $\rho_{gr}$, each coated with a layer of thickness $h$ of liquid of viscosity $\mu$, having a diameter $x$ and approach velocity $V_{app}$. The parameter which determines whether coalesence will occur is a Stokes number $Stk$:

$$Stk = \frac{\rho_{gr} V_{app} x}{16\mu} \tag{11.4}$$

This Stokes number is different from that used in cyclone scale-up in Chapter 7. The cyclone scale-up Stokes number $Stk_{50}$ incorporates the dimensionless ratio of particle size to cyclone diameter, i.e.:

$$Stk_{50} = \frac{\rho v x_{50}^2}{18\mu D} = \left[\frac{x_{50}}{D}\right]\frac{\rho v x_{50}}{18\mu} \quad \text{(where } v \text{ is the characteristic gas velocity)}$$

The Stokes number is a measure of the ratio of collisional kinetic energy to energy dissipated through viscous dissipation. For coalescence to occur the Stokes number must be less than a critical value $Stk^*$, given by

$$\text{Stk}^* = \left[1 + \frac{1}{e}\right]\ln\left(\frac{h}{h_a}\right) \tag{11.5}$$

where $e$ is the coefficient of restitution for the collision and $h_a$ is a measure of surface roughness of the granule.

Based on this criterion, three regimes of granule growth are identified for batch systems with relatively low agitation intensity. These are the non-inertial, inertial and coating regimes. Within the granulator at any time there will be a distribution of granule sizes and velocities which gives rise to a distribution of Stokes numbers for collisions. In the non-inertial regime Stokes number is less than $Stk^*$ for all granules and primary particles and practically all collisions result in coalescence. In this regime therefore the growth rate is largely independent of liquid viscosity, granule or primary particle size and kinetic energy of collision. The rate of wetting of the particles controls the rate of growth in this regime.

As granules grow some collisions will occur for which the Stokes number exceeds the critical value. We now enter the inertial regime in which the rate of growth is dependent on liquid viscosity, granule size and collision energy. The proportion of collisions for which the Stokes number exceeds the critical value increases throughout this regime and the proportion of successful collisions decreases. Once the average Stokes number for the powder mass in the granulator is comparable with the critical value, granule growth is balanced by breakage and growth continues by coating of primary particles onto existing granules, since these are the only possible successful collisions according to our criterion. This simple analysis breaks down when granule deformation cannot be ignored, as in high agitation intensity systems (see Ennis and Litster, 1997).

These regimes are equivalent to the growth mechanisms of nucleation, coalescence and layering observed by Sastry and Fuerstenau (1977).

*Granule consolidation*

Consolidation is the term used to describe the increase in granule density caused by closer packing of primary particles as liquid is squeezed out as a result of collisions. Consolidation can only occur whilst the binder is still liquid. Consolidation determines the porosity and density of the final granules. Factors influencing the rate and degree of consolidation include particle size, size distribution and binder viscosity.

*Granule breakage*

Granule breakage also takes place during the granulation process. Breakage mechanisms include shattering, fragmentation and wear. Shattering is the complete breakup of primary granules to their component primary particles. Fragmentation is the fracture of a granule to form two or more pieces. Wear (also called attrition or erosion) is the reduction in size of a granule by loss of primary particles from its surface. As granules collide partial transfer of granulated material from one granule to the other may also occur; this is referred to as abrasive transfer. These mechanisms were identified by Sastry and Fuerstenau (1977) and are summarized in Figure 11.3. Empirical and theoretical approaches exist for modelling the different breakage mechanisms (see Ennis and Litster, 1997). In practice breakage may be controlled by altering the granule properties (e.g. increase fracture toughness and increase resistance to attrition) and by making changes to the process (e.g. reduce agitation intensity).

## 11.3.3 Simulation of the Granulation Process

As in the processes of comminution and crystallization, the simulation of granulation hinges on the population balance. The population balance tracks the size distribution (by number, volume or mass) with time as the process progresses. It is a statement of the material balance for the process at a given instant. In the case of granulation the instantaneous population balance equation is often written in terms of the number distribution of the volume of granules, $n(v, t)$ (rather than granule diameter since volume is assumed to be conserved in any coalescence). $n(v, t)$ is the number frequency distribution of granule volume at time $t$. Its units are number of granules per unit volume size increment, per unit volume of granulator. In words it is written as (Equation (11.6)):

$$
\underbrace{\begin{bmatrix} \text{rate of} \\ \text{increase} \\ \text{of number} \\ \text{of granules} \\ \text{in size} \\ \text{interval} \\ v \text{ to } v + dv \end{bmatrix}}_{1} = \underbrace{\begin{bmatrix} \text{rate of} \\ \text{inflow} \\ \text{of granules} \\ \text{in size} \\ \text{interval} \\ v \text{ to } v + dv \end{bmatrix}}_{2} - \underbrace{\begin{bmatrix} \text{rate of} \\ \text{outflow} \\ \text{of granules} \\ \text{in size} \\ \text{interval} \\ v \text{ to } v + dv \end{bmatrix}}_{3} + \underbrace{\begin{bmatrix} \text{rate at which} \\ \text{granules} \\ \text{enter size} \\ \text{range} \\ v \text{ to } v + dv \\ \text{by growth} \end{bmatrix}}_{4} - \underbrace{\begin{bmatrix} \text{rate at} \\ \text{which} \\ \text{granules} \\ \text{leave size} \\ \text{range} \\ v \text{ to } v + dv \\ \text{by breakage} \end{bmatrix}}_{5}
$$

$$(11.6)$$

Term 4 may be expanded to account for the different growth mechanisms and term 5 may be expanded to include the different mechanisms by which breakage occurs.

For a constant volume granulator the terms in the population balance equation become:

*Term 1*

$$\frac{\partial n(v, t)}{\partial t} \tag{11.7}$$

*Term 2–Term 3*

$$\frac{Q_{in}}{V} n_{in}(v) - \frac{Q_{out}}{V} n_{out}(v) \tag{11.8}$$

*Term 4*

$$\begin{bmatrix} \text{net rate of} \\ \text{growth} \\ \text{by coating} \end{bmatrix} + \begin{bmatrix} \text{rate of growth} \\ \text{by nucleation} \end{bmatrix} + \begin{bmatrix} \text{rate of growth} \\ \text{by coalescence} \end{bmatrix}$$

Growth by coating causes granules to grow into and out of the size range $v$ to $v + dv$.

$$\text{Hence, the net rate of growth by coating} = \frac{\partial G(v)n(v, t)}{\partial v} \tag{11.9}$$

$$\text{The rate of growth by nucleation} = B_{nuc}(v) \tag{11.10}$$

$G(v)$ is the volumetric growth rate constant for coating, $B_{nuc}(v)$ is the rate constant for nucleation. It is often acceptable to assume that $G(v)$ is proportional to the available granule surface area; this is equivalent to assuming a constant linear growth rate, $G(x)$. In reference to granulation processes, the use of the term 'nucleation' is sometimes confusing. Sastry and Loftus (1989) suggest that in a granulation process there is in general a continuous phase and a particulate phase. The continuous phase may, for example, be a solution, slurry or very small particles; the particulate phase is made up of original particles and granules. Nuclei are formed from the continuous phase and then become part of the particulate phase. What constitutes the continuous phase depends on the nature of the granulation, and the cut-off between continuous and particulate phases may be arbitrary. It will be appreciated therefore that the form of the granulation rate constant may vary considerably depending in the definitions of continuous and particulate phases.

The rate of growth of granules by coalescence may be written as (Randolph and Larson, 1971)

$$\underbrace{\left\{\frac{1}{2}\int_0^v \beta(u, v - u, t)\, n(u, t)\, n(v - u, t)\, du\right\}}_{\text{term (i)}} + \underbrace{\left\{\int_0^\infty \beta(u, v, t)\, n(u, t)\, n(v, t)\, du\right\}}_{\text{term (ii)}}$$

(11.11a)

Term (i) is the rate of formation of granules of size $v$ by coalescence of smaller granules. Term (ii) is the rate at which granules of size $v$ are lost by coalescence to form larger granules. $\beta$ is called the coalescence kernel. The rate of coalescence of two granules of volume $u$ and $(v - u)$ to form a new granule of volume $v$ is assumed to be directly proportional to the product of the number densities of the starting granules:

$$\begin{bmatrix}\text{collision rate of granules} \\ \text{of volume } u \text{ and } (v - u)\end{bmatrix} \propto \begin{bmatrix}\text{number density of} \\ \text{granules of volume } u\end{bmatrix} \times \begin{bmatrix}\text{number density of} \\ \text{granules of volume} \\ (v - u)\end{bmatrix}$$

The number densities are time dependent in general and so $n(u, t)$, the number density of granules of volume $u$ at time $t$ and $n(v - u, t)$ is the number density of granules of volume $(v - u)$ at time $t$. The constant in this proportionality is $\beta$, the coalescence kernel or coalescence rate constant, which is in general assumed to be dependent on the volumes of the colliding granules. Hence, $\beta(u, v - u, t)$ is the coalescence rate constant for collision between granules of volume $u$ and $(u - v)$ at time $t$.

The above assumes a pseudo-second order process of coalescence in which all granules have an equal opportunity to collide with all other particles. In real granulation systems this assumption does not hold and collision opportunities are limited to local granules. Sastry and Fuerstenau (1970) suggested that for a batch granulation system, which was effectively restricted in space, the appropriate form for terms (i) and (ii) was

$$\underbrace{\left\{\frac{1}{2N(t)}\int_0^v \beta(u, v - u, t)n(u, t)n(v - u, t)\, du\right\}}_{\text{term (i)}} + \underbrace{\left\{\frac{1}{N(t)}\int_0^\infty \beta(u, v, t)n(u, t)n(v, t)\, du\right\}}_{\text{term (ii)}}$$

(11.11b)

Where $N(t)$ is the total number of granules in the system at time $t$.

The integrals in Equation (11.11) account for all the possible collisions and the $\frac{1}{2}$ in term (i) of this equation ensures that collisions are only counted once.

In practice coalescence kernels are determined empirically and based on laboratory or plant data specific to the granulation process and the product.

According to Sastry (1975) the coalescence kernel is best expressed in two parts:

$$\beta(u, v) = \beta_0 \beta_1(u, v) \tag{11.12}$$

$\beta_0$ is the coalescence rate constant which determines the rate at which successful collisions occur and hence governs average granule size. It depends on solid and liquid properties and agitation intensity. $\beta_1(u, v)$ governs the functional dependency of the kernel on the sizes of the coalescing granules, $u$ and $v$. $\beta_1(u, v)$ determines the shape of the size distribution of granules. Various forms of $\beta$ have been published; Ennis and Litster (1997) suggest the form shown in Equation (11.13), which is consistent with the granulation regime analysis described above:

$$\beta(u, v) = \begin{cases} \beta_0, & w < w^* \\ 0, & w > w^* \end{cases} \quad \text{where } w = \frac{(uv)^a}{(u+v)^b} \tag{11.13}$$

$w^*$ is the critical average granule volume in a collision and corresponds to the critical Stokes number value $Stk^*$. From the definition the Stokes number given in Equation (11.5), the critical diameter $x^*$ would be

$$x^* = \frac{16\mu Stk^*}{\rho_{gr} V_{app}} \tag{11.14}$$

and so, assuming spherical granules,

$$w^* = \frac{\pi}{6} \left[ \frac{16\mu Stk^*}{\rho V_{app}} \right]^3 \tag{11.15}$$

The exponents $a$ and $\beta$ are dependent on granule deformability and on the granule volumes $u$ and $v$. In the case of small feed particles in the non-inertial regime, $\beta$ reduces to the size-independent rate constant $\beta_0$ and the coalescence rate is independent of granule size. Under these conditions the mean granule size increases exponentially with time. Coalescence stops ($\beta = 0$) when the critical Stokes number is reached.

Using this approach, Adetayo and Ennis (1997) were able to demonstrate the three regimes of granulation (nucleation, transition and coating) traditionally observed in drum granulation and to model a variety of apparently contradictory observations.

*Term 5*

$$\begin{bmatrix} \text{rate of breakage} \\ \text{by attrition} \end{bmatrix} + \begin{bmatrix} \text{rate of breakage} \\ \text{by fragmentation} \end{bmatrix}$$

$$\text{rate of breakage by attrition} = \frac{\partial A(v)n(v, t)}{\partial v} \tag{11.16}$$

The rate of attrition depends on the material to be granulated, the binder, the type of equipment used for granulation and the intensity of agitation. The rate of breakage may also be accounted for by the use of selection and breakage function as used in the simulation of population balances in comminution (see Chapter 10). $A(v)$ is the rate constant for attrition.

## 11.3.4   Granulation Equipment

Three categories of granulator are in common use: tumbling, mixer and fluidized granulators. Typical product properties, scale and applications of each of these granulators are summarized in Table 11.1 (after Ennis and Litster, 1997).

### Tumbling granulators

In tumbling granulators a tumbling motion is imparted to the particles in an inclined cylinder (drum granulator) or pan (disc granulator, Figure 11.4). Tumbling granulators operate in continuous mode and are able to deal with large throughputs (see Table 11.1). Solids and liquid feeds are delivered continuously to the granulator. In the case of the disc granulator the tumbling action gives rise to a natural classification of the contents according to size. Advantage is taken of this effect and the result is a product with a narrow size distribution.

### Mixer granulators

In mixer granulators the motion of the particles is brought about by some form of agitator rotating at low or high speed on a vertical or horizontal axis. Rotation speeds vary from 50 revolutions per minute in the case of the horizontal pug mixers used for fertilizer granulation to over 3000 revolutions per minute in the case of the vertical Schugi high shear continuous granulator used for detergents and agricultural chemicals.

### Fluidized bed granulators

In fluidized bed granulators the particles are set in motion by fluidizing air. The fluidized bed may be either a bubbling or spouted bed (see Chapter 5) and may operate in batch or continuous mode. Liquid binder and wetting agents are sprayed in atomized form above or within the bed. The advantages which this granulator has over others include good heat and mass transfer, mechanical simplicity, ability to combine the drying stage with the granulation stage and ability to produce small granules from powder feeds. On the other hand

**Table 11.1** Types of granulator, their features and applications (after Ennis and Litster, 1997)

| Method | Product granule size (mm) | Granule density | Scale/throughput | Comments | Typical applications |
|---|---|---|---|---|---|
| Tumbling (disc, drum) | 0.5–20 | Moderate | 0.5–800 tonnes/h | Very spherical granules | Fertilizers, iron ore, agricultural chemicals |
| Mixer (continuous and batch high shear) | 0.1–2 | Low | <50 tonnes/h | Handles cohesive materials well | Chemicals, detergents, pharmaceuticals, ceramics |
| Fluidized (bubbling beds, spouted beds) | 0.1–2 | High | <500 kg batch | Good for coating, easy to scale up | Continuous (fertilizers, detergents) batch (pharmaceuticals, agricultural chemicals) |

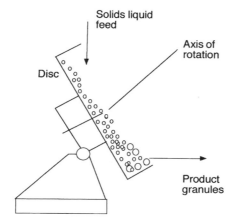

**Figure 11.4**   Schematic of disc granulator

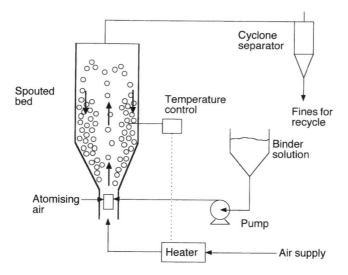

**Figure 11.5**   Schematic of a spouted fluidized bed granulator circuit (Liu and Litster, 1993)

running costs and attrition rates can be high compared with other devices. A typical spouted bed granulator circuit is shown in Figure 11.5.

For further details of industrial equipment for granulation and other means of size enlargement the reader is referred to Ennis and Litster (1997) or Capes (1980).

# 12

# Fire and Explosion Hazards of Fine Powders

## 12.1 INTRODUCTION

Finely divided combustible solids, or dusts, dispersed in air can give rise to explosions in much the same way as flammable gases. In the case of flammable gases, fuel concentration, local heat transfer conditions, oxygen concentration and initial temperature all affect ignition and resulting explosion characteristics. In the case of dusts, however, more variables are involved (e.g. particle size distribution, moisture content) and so the analysis and prediction of dust explosion characteristics is more complex than for the flammable gases. Dust explosions have been known to give rise to serious property damage and loss of life. Most people are probably aware that dust explosions have occurred in grain silos, flour mills and in the processing of coal. However, explosions of dispersions of fine particles of metals (e.g. aluminium), plastics, sugar and pharmaceutical products can be particularly potent. Process industries where fine combustible powders are used and where particular attention must be directed towards control of dust explosion hazard include: plastics, food processing, metal processing, pharmaceuticals, agricultural, chemicals and coal. Process steps where fine powders are heated have a strong association with dust explosion; examples include dilute pneumatic conveying and spray drying, which involves heat and a dilute suspension.

In this chapter, the basics of combustion are outlined followed by the fundamentals specific to dust explosions. The measurement and application of dust explosion characteristics, such as ignition temperature, range of flammable concentrations minimum ignition energy are covered. Finally the available methods for control of dust explosion hazard are discussed.

## 12.2  COMBUSTION FUNDAMENTALS

### 12.2.1  Flames

A flame is a gas rendered luminous by emission of energy produced by chemical reaction. In a stationary flame (for example a candle flame or gas stove flame) unburned fuel and air flow into the flame front as combustion products flow away from the flamefront. A stationary flame may be from either premixed fuel and air, as observed in a Bunsen burner with the air hole open, or by diffusion of air into the combustion zone, as for a Bunsen burner with the air hole closed.

When the flame front is not stationary it is called an explosion flame. In this case the flame front passes through a homogeneous premixed fuel–air mixture. The heat released and gases generated result in either an uncontrolled expansion effect or, if the expansion is restricted, a rapid build-up of pressure.

### 12.2.2  Explosions and Detonations

Explosion flames travel through the fuel–air mixture at velocities ranging from a few metres per second to several hundreds of metres per second and this type of explosion is called a deflagration. Flame speeds are governed by many factors including the heat of combustion of the fuel, the degree of turbulence in the mixture and the amount of energy supplied to cause ignition. It is possible for flames to reach supersonic velocities under some circumstances. Such explosions are accompanied by pressure shock waves, are far more destructive and called detonations. The increased velocities result from increased gas densities generated by pressure waves. It is not yet understood what conditions give rise to detonations. However, in practice it is likely that all detonations begin as deflagrations.

### 12.2.3  Ignition, Ignition Energy, Ignition Temperature – a Simple Analysis

Ignition is the self-propagation of a combustion reaction through a fuel–air mixture after the initial supply of energy. Ignition of fuel–air mixture can be analysed in a manner similar to that used for runaway reactions (thermal explosions). Consider an element of fuel air mixture of volume $V$ and surface area $A$, in which the volumetric concentration of fuel is $C$. If the temperature of the fuel–air mixture in the element is $T_i$ and if the rate at which heat is lost from the element to the surroundings (at temperature $T_s$) is governed by a heat transfer coefficient $h$, then the rate of heat loss to surroundings, $Q_s$ is

$$Q_s = hA(T_i - T_s) \tag{12.1}$$

The variation of the combustion reaction rate with temperature will be governed by the Arrhenius equation. For a reaction which is first order in fuel concentration,

$$V\frac{dC}{dt} = C\rho_{mfuel} Z \exp\left[-\frac{E}{RT}\right] \tag{12.2}$$

where $Z$ is the pre-exponential coefficient, $E$ is the reaction activation energy, $R$ is the ideal gas constant and $\rho_{mfuel}$ is the molar density of the fuel.

The rate $Q_a$ at which heat is absorbed by the fuel–air mixture in the element is

$$Q_a = V\frac{dT}{dt}[C\rho_{m_{fuel}} C_{P_{fuel}} + (1 - C)\rho_{m_{air}} C_{P_{air}} \tag{12.3}$$

where $C_{P_{fuel}}$ and $C_{P_{air}}$ are the specific heat capacities of the fuel and air and $\rho_{fuel}$ and $\rho_{m_{air}}$ are the molar densities of the fuel and air respectively.

If $Q_{input}$ is the rate at which heat energy is fed into the element from outside, then the heat balance for the element becomes

$$\underset{(1)}{Q_{input}} + \underset{(2)}{(-\Delta H)C\rho_{m_{fuel}} Z \exp\left[-\frac{E}{RT_R}\right]}$$

$$= V\frac{dT_R}{dt}\underset{(3)}{[C\rho_{m_{fuel}} C_{P_{fuel}} + (1 - C)\rho_{mair} C_{P_{air}}]} + \underset{(4)}{hA(T_{Ri} - T_s)} \tag{12.4}$$

It is instructive to analyse this heat balance graphically for the steady-state condition (term 3 is zero). We do this by plotting the rates of heat loss to the surroundings and the rate of heat generation by the combustion reaction as a function of temperature. The former of course results in a straight line of slope $hA$ and intercept of the temperature axis $T_s$. The rate of heat generation by reaction within the element is given by term 2 and results in an exponential curve. A typical plot is shown in Figure 12.1. Analysing this type of plot gives us insight into the meaning of ignition, ignition energy, ignition temperature etc.

Consider initially the case where $Q_{input}$ is zero. Referring to the case shown in Figure 12.1, we see that at an initial element temperature $T_i$ the rate of heat loss from the element is greater than the rate of heat generation and so the temperature of the element will decrease until point A is reached. Any initial temperature between $T_B$ and $T_A$ will result in the element cooling to $T_A$. This is a stable condition.

If, however, the initial temperature is greater than $T_B$, the rate of heat generation within the element will be always greater than the rate of heat loss to the surroundings and so the element temperature will rise, exponentially. Thus initial temperatures beyond $T_B$ give rise to an unstable condition. $T_B$ is the *ignition temperature*, $T_{ig}$ for the fuel–air mixture in the element. *Ignition energy* is the energy that we must supply from the outside in order to raise the

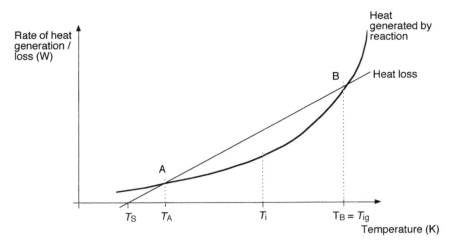

**Figure 12.1**  Variation of rates of heat generation and loss with temperature of element

mixture from its initial temperature $T_i$ to the ignition temperature $T_{ig}$. Since the element is continuously losing energy to the surroundings, the ignition energy will actually be a rate of energy input, $Q_{input}$. This raises the heat generation curve by an amount $Q_{input}$, reducing the value of $T_{ig}$ (Figure 12.2). The conditions for heat transfer from the element to the surroundings are obviously important in determining the ignition temperature and energy. There are cases where the heat loss curve will be always lower than the heat generation curve (Figure 12.3). Under such circumstances the mixture may self-ignite; this is referred to as auto-ignition or spontaneous ignition.

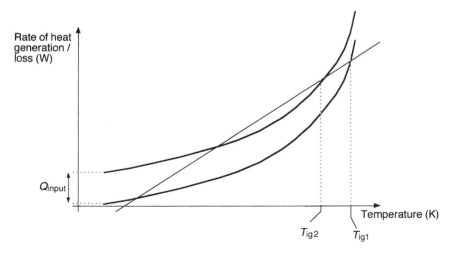

**Figure 12.2**  Variation in rates of heat generation and loss with temperature of element; the effect on ignition temperature of adding energy

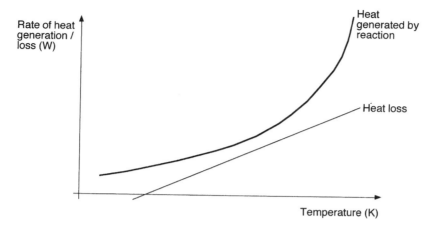

**Figure 12.3** Variation in rates of heat generation and loss with temperature of element; rate of heat generation always greater that rate of heat loss.

In many combustion systems there is an appreciable interval (from milliseconds to several minutes between arrival at the ignition temperature and the apparent onset of ignition. This is known as the *ignition delay*. It is not well understood, but based on the analysis above it is related to the time required for the reaction in the element to go to completion once the ignition temperature has been reached.

If there is adequate oxygen for combustion and the fuel concentration in the element is increased the rate of heat generation by combustion will also increase. Whether the ignition temperature and energy will be affected depends on the relative physical properties (specific heat capacity and conductivity) of the fuel and air.

## 12.2.4 Flammability Limits

From the above analysis it can be seen that below a certain fuel concentration ignition will not occur, since the rate of heat generation within the element is insufficient to match the rate of heat loss to the surroundings ($T_{ig}$ is never reached). This concentration is known as the lower flammability limit $C_{fL}$ of the fuel–air mixture. It is generally measured under standard conditions in order to give reproducible heat transfer conditions.

There is also an upper flammability limit, $C_{fU}$. This is the concentration of fuel in the fuel–air mixture above which a flame cannot be propagated. Interpreted in terms of the above analysis, this is the temperature above which the physical properties of the fuel–air mixture are unfavourable, i.e. the temperature rise for a given heat release is too low or the rate of heat transfer to the surroundings is too high to achieve ignition.

Thus, in general, there is a range of fuel concentration in air within which a flame can be propagated. From the analysis above it will be apparent that this range will widen ($C_{fL}$ will decrease and $C_{fU}$ will increase) as the initial temperature of the mixture is increased. In practice, therefore, flammability limits are measured and quoted at standard temperatures (usually 20°C).

## Minimum oxygen for combustion

At the lower limit of flammability there is more oxygen available than is required for stoichiometric combustion of the fuel. For example, the lower flammability limit for propane in air at 20°C is 2.2% by volume.
 For complete combustion of propane according to the reaction

$$C_3H_8 + 5O_2 \rightarrow 3CO_2 + 4H_2O$$

Five volumes of oxygen are required per volume of fuel propane.
 In a fuel–air mixture with 2.2% propane the ratio of air to propane is

$$\frac{100 - 2.2}{2.2} = 44.45$$

and since air is approximately 21% oxygen, the ratio of oxygen to propane is 9.33. Thus in the case of propane at the lower flammability limit oxygen is in excess by approximately 87%.
 It is therefore possible to reduce the concentration of oxygen in the fuel–air mixture whilst still maintaining the ability to propagate a flame. If the oxygen is replaced by a gas which has similar physical properties (nitrogen for example) the effect on the ability of the mixture to maintain a flame is minimal until the stoichiometric ratio of oxygen to fuel is reached. The oxygen concentration in the mixture under these conditions is known as the minimum oxygen for combustion (MOC). Minimum oxygen for combustion is therefore the stoichiometric oxygen equivalent to the lower flammability limit. Thus

$$MOC = C_{fL} \times \left[\frac{\text{moles } O_2}{\text{moles fuel}}\right]_{\text{stoich}}$$

For example, for propane, since under stoichiometric conditions five volumes of oxygen are required per volume of fuel propane,

$$MOC = 2.2 \times 5 = 11\% \text{ oxygen by volume}$$

## 12.3   COMBUSTION IN DUST CLOUDS

### 12.3.1   Fundamental Specific to Dust Cloud Explosions

The combustion rate of a solid in air will in most cases be limited by the surface area of solid presented to the air. Even if the combustible solid is in particulate form a few millimetres in size, this will still be the case. However, if the particles of solid are small enough to be dispersed in air without too much propensity to settle, the reaction rate will be great enough to permit an explosion flame to propagate. For a dust explosion to occur the solid material of which the particles are composed must be combustible, i.e. it must react exothermically with the oxygen in air. However, not all combustible solids give rise to dust explosions.

To make our combustion fundamentals considered above applicable to dust explosions we need only to add in the influence of particle size on reaction rate. Assuming the combustion reaction rate is now determined by the surface area of solid fuel particles (assumed spherical) exposed to the air, the heat release term 2 in Equation (12.4) becomes:

$$(-\Delta H)\left(\frac{6C}{x\rho_{m_{fuel}}}\right) Z \exp\left[-\frac{E}{RT}\right]\tag{12.5}$$

where $x$ is the particle size and $\rho_{m_{fuel}}$ is the molar density of the solids fuel.

The rate of heat generation by the combustion reaction is therefore inversely proportional to the dust particle size. Thus, the likelihood of flame propagation and explosion will increase with decreasing particle size. Qualitatively, this is because finer fuel particles;

- are more readily form a dispersion in air;
- have a larger surface area per unit mass of fuel;
- offer a greater surface area for reaction (higher reaction rate, as limiting);
- consequently generate more heat per unit mass of fuel;
- have a greater heat-up rate.

### 12.3.2   Characteristics of Dust Explosions

Consider design engineers wishing to know the potential fire and explosion hazards associated with a particulate solid made or used in a plant which they are designing. They are faced with the same problem which they face in gathering any 'property' data of particulate solids; unlike liquids and gases there is little published data, and what is available is unlikely to be relevant. The particle size distribution, surface properties and moisture content all influence the potential fire hazard of the powder, so unless the engineers can be sure that their powder is identical in every way to the powder used for the published data, they must have the explosion characteristics of the powder tested. Having made the decision, the engineers must ensure that the sample given to the test laboratory is truly representative of the material to be produced or used in the final plant.

Although there has been recent progress towards uniform international testing standards for combustible powders, there remain some differences. However, most tests include an assessment of the following explosion characteristics:

- minimum dust concentration for explosion;

- minimum energy for ignition;

- minimum ignition temperature;

- maximum explosion pressure;

- maximum rate of pressure rise during explosion;

- minimum oxygen for combustion.

An additional classification test is sometimes used. This is simply a test for explosibility in the test apparatus, classifying the dust as able or unable to ignite and propagate a flame in air at room temperature under test conditions. This test in itself is not very useful particularly if the test conditions differ significantly from the plant conditions.

### 12.3.3  Apparatus for Determination of Dust Explosion Characteristics

There are several different devices for determination of dust explosion characteristics. All devices include a vessel which may be open or closed, an ignition source which may be electrical spark or electrically heated wire coil and a supply of air for dispersion of the dust. The simplest apparatus is known as the vertical tube apparatus and is shown schematically in Figure 12.4. The sample dust is placed in the dispersion cup. Delivery of dispersion air to the cup is via a solenoid valve. Ignition may be either by electrical spark across electrodes or by heated coil. The vertical tube apparatus is used for the classification test and for determination of minimum dust concentration for explosion, minimum energy for ignition and in a modified form for minimum oxygen for combustion.

A second apparatus known as the 20 litre sphere is used for determination of maximum explosion pressure and maximum rate of pressure rise during explosion. These give an indication of the severity of explosion and enable the design of explosion protection equipment. This apparatus, which is shown schematically in Figure 12.5, is based on a spherical 20 litre pressure vessel fitted with a pressure transducer. The dust to be tested is first charged

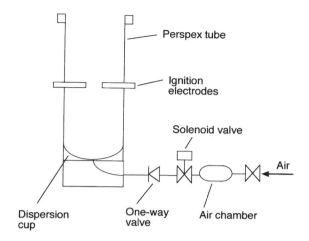

**Figure 12.4**  Vertical tube apparatus for determination of dust explosion characteristics

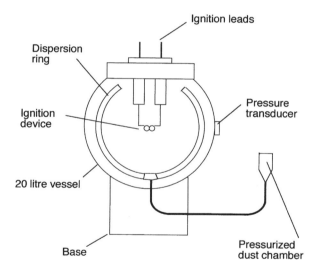

**Figure 12.5**  20 litre sphere apparatus for determination of dust explosion characteristics (after Lunn, 1992)

to a reservoir and then blown by air into the sphere via a perforated dispersion ring. The vessel pressure is reduced to about 0.4 bar before the test so that upon injection of the dust, the pressure rises to atmospheric. Ignition is by a pyrotechnical device with a standard total energy (typically 10 kJ) positioned at the centre of the sphere. The delay between dispersion of the dust and initiation of the ignition source has been found to affect the results. Turbulence caused by the air injection influences the rate of the combustion reaction. A standard delay of typically 60 ms is therefore employed in order to ensure the reproducibility of the test. There is also a 1 m³ version of this apparatus.

The third basic test device is the Godbert–Greenwald furnace apparatus, which is used to determine the minimum ignition temperature and the explosion characteristics at elevated temperatures. The apparatus includes a vertical electrically heated furnace tube which can be raised to controlled temperatures up to 1000°C. The dust under test is charged to a reservoir and then dispersed through the tube. If ignition occurs the furnace temperature is lowered in 10°C steps until ignition does not occur. The lowest temperature at which ignition occurs is taken as the ignition temperature. Since the quantity of dust used and the pressure of the dispersion air both affect the result, these are varied to obtain a minimum ignition temperature.

## 12.3.4   Application of the Test Results

The *minimum dust concentration for explosion* is measured in the vertical tube apparatus and is used to give an indication of the quantities of air to be used in extraction systems for combustible dusts. Since dust concentrations can vary widely with time and location in a plant it is not considered wise to use concentration control as the sole method of protection against dust explosion.

The *minimum energy for ignition* is measured primarily to determine whether the dust cloud could be ignited by an electrostatic spark. Ignition energies of dusts can be as low as 15 mJ; this quantity of energy can be supplied by an electrostatic discharge.

The *minimum ignition temperature* indicates the maximum temperature for equipment surfaces in contact with the powder. For new materials it also permits comparison with well-known dusts for design purposes. Table 12.1 gives some values of explosion parameters for common materials.

The *maximum explosion pressure* is usually in the range 8–13 bar and is used mainly to determine the design pressure for equipment when explosion containment or protection is opted for as the method of dust explosion control.

The *maximum rate of pressure rise* during explosion is used in the design of explosion relief. It has been demonstrated that the maximum rate of pressure

**Table 12.1**   Explosion parameters for some common materials (Schofield, 1984)

| Dust | Mean particle size (μm) | Maximum explosion pressure (bar) | Maximum rate of pressure rise (bar/s) | $K_{St}$ |
|------|------|------|------|------|
| Aluminium | 17 | 7.0 | 572 | 155 |
| Polyester | 30 | 6.1 | 313 | 85 |
| Polyethylene | 14 | 5.9 | 494 | 134 |
| Wheat | 22 | 6.1 | 239 | 65 |
| Zinc | 17 | 4.7 | 131 | 35 |

rise in a dust explosion is inversely proportional to the cube root of the vessel volume, i.e.

$$\left(\frac{\mathrm{d}p}{\mathrm{d}t}\right)_{\max} = V^{1/3} K_{St} \qquad (12.6)$$

The value of $K_{St}$ is found to be constant for a given powder. Typical values are given in Table 12.1. The severity of dust explosions is classified according to the St Class based on the $K_{St}$ value (see Table 12.2).

The *minimum oxygen for combustion* (MOC) is used to determine the maximum permissible oxygen concentration when inerting is selected as the means of controlling the dust explosion. Organic dusts have an MOC of about 11% if nitrogen is the diluent and 13% in the case of carbon dioxide. Inerting requirements for metal dusts are more stringent since MOC values for metals can be far lower.

## 12.4   CONTROL OF THE HAZARD

### 12.4.1   Introduction

As with the control of any process hazard, there is a hierarchy of approaches that can be taken to control dust explosion hazard. These range from the most desirable strategic approach of changing the process to eliminate the hazardous powder altogether to the merely tactical approach of avoiding ignition sources. In approximate order of decreasing strategic component, the main approaches are listed below:

- Change the process to eliminate the dust.
- Design the plant to withstand the pressure generated by any explosion.
- Remove the oxygen by complete inerting.
- Reduce oxygen to below MOC.
- Add moisture to the dust.
- Add diluent powder to the dust.

**Table 12.2**   Dust explosion classes based on 1 m³ test apparatus

| Dust explosion class | $K_{St}$ (bar m/s) | Comments |
| --- | --- | --- |
| St 0 | 0 | Non-explosible |
| St 1 | 0–200 | Weak to moderately explosible |
| St 2 | 200–300 | Strongly explosible |
| St 3 | > 300 | Very strongly explosible |

- Detect start of explosion and inject suppressant.

- Vent the vessel to relieve pressure generated by the explosion.

- Control dust concentration to be outside flammability limits.

- Minimise dust cloud formation.

- Exclude ignition sources.

## 12.4.2   Ignition Sources

Excluding ignition sources sounds a sensible policy. However, statistics of dust explosions indicate that in a large proportion of incidents the source of ignition was unknown. Thus, whilst it is good policy to avoid sources of ignition as far as possible, this should not be relied on as the sole protection mechanism. It is interesting to look briefly at the ignition sources which have been associated with dust explosions.

- *Flames*. Flames from the burning of gases, liquids or solids are effective sources of ignition for flammable dust clouds. Several sources of flames can be found in process plant during normal operation (e.g. burners, pilot flames etc.) and during maintenance (e.g. welding and cutting flames). These flames would usually be external to the vessels and equipment containing the dust. To avoid exposure of dust clouds to flames, therefore, good house-keeping is required to avoid a build-up of dust which may generate a cloud and a good permit-to-work system should be in place to ensure a safe environ-ment before maintenance commences.

- *Hot surfaces*. Careful design is required to ensure that surfaces likely to be in contact with dust do not reach temperatures which can cause ignition. Attention to detail is important; for example ledges inside equipment should be avoided to prevent settling of dust and possible self ignition. Dust must not be able to build up on hot or heated surfaces, otherwise surface temperatures will rise as heat dissipation from the surface is reduced. Outside the vessel care must also be taken; for example if dust is allowed to settle on electric motor housings, overheating and ignition may occur.

- *Electric sparks*. Sparks produced in the normal operation of electrical power sources (by switches, contact breakers and electric motors) can ignite dust clouds. Special electrical equipment is available for application in areas where there is a potential for dust explosion hazard. Sparks from electrostatic discharges are also able to ignite dust clouds. Electrostatic charges are developed in many processing operations (particularly those involving dry powders) and so care must be taken to ensure that such charges are led to earth to prevent accumulation and eventual discharge. Even the energy in the charge developed on a process operator can be sufficient to ignite a dust cloud.

- *Mechanical sparks and friction.* Sparks and local heating caused by friction or impact between two metal surfaces or between a metal surface and foreign objects inadvertently introduced into the plant have been known to ignite dust clouds.

### 12.4.3   Venting

If a dust explosion occurs in a closed vessel at one atmosphere, the pressure will rise rapidly (up to and sometimes beyond 600 bar/s) to maximum .of around 10 bar. If the vessel is not designed to withstand such a pressure, deformation and possible rupture will occur. The principle of explosion venting is to discharge the vessel contents through an opening or vent to prevent the pressure rising above the vessel design pressure. Venting is a relatively simple and inexpensive method of dust explosion control but cannot be used when the dust, gas or combustion products are toxic or in some other way hazardous, or when the rate of pressure rise is greater than 600 bar/s (Lunn, 1992). The design of vents is best left to the expert although there are published guides (Lunn, 1992). The mass and type of the vent determine the pressure at which the vent opens and the delay before it is fully open. These factors together with the size of the vent determine the rate of pressure rise and the maximum pressure reached after the vent opens. Figure 12.6 shows typical pressure rise profiles for explosions in a vessel without venting and with vents of different size.

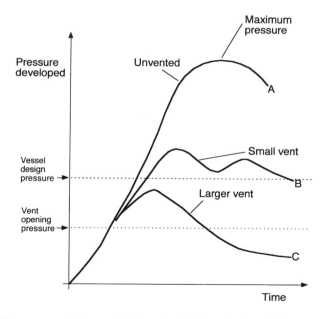

**Figure 12.6**   Pressure variation with time for dust explosions (A) unvented, (B) vented with inadequate vent area and (C) vented with adequate vent area (after Schofield, 1985)

## 12.4.4   Suppression

The pressure rise accompanying a dust explosion is rapid but it can be detected in time to initiate some action to suppress the explosion. Suppression involves discharging a quantity of inert gas or powder into the vessel in which the explosion has commenced. Modern suppression systems triggered by the pressure rise accompanying the start of the explosion have response times of the order of a few milliseconds and are able to effectively extinguish the explosion. The fast acting trigger device can also be used to vent the explosion, isolate the plant item or shut down the plant if necessary.

## 12.4.5   Inerting

Nitrogen and carbon dioxide are commonly used to reduce the oxygen concentration of air to below the minimum oxygen for combustion (MOC). Even if the oxygen concentration is not reduced as far as the MOC value, the maximum explosion pressure and the maximum rate of pressure rise are much reduced (Palmer 1990). Total replacement of oxygen is a more expensive option, but provides an added degree of safety.

## 12.4.6   Minimize Dust Cloud Formation

In itself this cannot be relied on as a control measure, but should be incorporated in the general design philosophy of plant involving flammable dusts. Examples are (1) use of dense phase conveying as an alternative to dilute phase, (2) use of cyclone separators and filters instead of settling vessels for separation of conveyed powder from air, (3) avoiding situations where a powder stream is allowed to fall freely through air (e.g. in charging a storage hopper). Outside the vessels of the process good housekeeping practice should ensure that deposits of powder are not allowed to build up on ledges and surfaces within a building. This avoids secondary dust explosions caused when these deposits are disturbed and dispersed by a primary explosion or shock wave.

## 12.4.7   Containment

Where plant vessels are of small dimensions it may be economic to design them to withstand the maximum pressure generated by the dust explosion (Schofield and Abbott, 1988). The vessel may be designed to contain the explosion and be replaced afterwards or to withstand the explosion and be reusable. In both cases design of the vessel and its accompanying connections and ductwork is a specialist task. For large vessels the cost of design and construction to contain dust explosions is usually prohibitive.

## 12.5 WORKED EXAMPLES

### WORKED EXAMPLE 12.1

It is proposed to protect a section of duct used for pneumatically transporting a plastic powder in air by adding a stream of nitrogen. The air flowrate in the present system is 1.6 m³/s and the air carries 3% powder by volume. If the minimum oxygen for combustion (by replacement of oxygen with nitrogen) of the powder is 11% by volume, what is the minimum flowrate of nitrogen which must be added to ensure safe operation?

*Solution*

The current total air flow of 1.6 m³/s includes 3% by volume of plastic powder and 97% air (made up of 21% oxygen and 79% nitrogen by volume). In this stream the flowrates are therefore:

powder:    0.048 m³/s

oxygen:    0.3259 m³/s

nitrogen:  1.226 m³/s.

At the limit, the final concentration of the flowing mixture should be 11% by volume. Hence, using a simple mass balance assuming constant densities,

$$\frac{\text{volume flow of } O_2}{\text{total volume flow}} = \frac{0.3259}{1.6 + n} = 0.11$$

from which, the minimum required flow rate of added nitrogen, $n = 1.36$ m³/s.

### WORKED EXAMPLE 12.2

A combustible dust has a lower flammability concentration limit in air at 20°C of 0.9% by volume. A dust extraction system operating at 2 m³/s is found to have a dust concentration of 2% by volume. What minimum flowrate of additional air must be introduced to ensure safe operation?

*Solution*

Assume that the dust explosion hazard will be reduced by bringing the dust concentration in the extract to below the lower flammability limit. In 2 m³/s of extract, the flow rates or air and dust are

air:    1.96 m³/s

dust:   0.04 m³/s

At the limit, the dust concentration after addition of dilution air will be 0.9%, hence:

$$\frac{\text{volume of dust}}{\text{total volume}} = \frac{0.04}{2+n} = 0.09$$

from which the minimum required flow rate of added dilution air, $n = 2.44 \text{ m}^3/\text{s}$.

## WORKED EXAMPLE 12.3

A flammable dust is suspended in air at a concentration within the flammable limits and with an oxygen concentration above the minimum oxygen for combustion. Sparks generated by a grinding wheel pass through the suspension at high speed, but no fire or explosion results. Explain why.

### Solution

In this case it is likely that the temperature of the sparks will be above the measured minimum ignition temperature and the energy available is greater than the minimum ignition energy. However, an explanation might be that the heat transfer conditions are unfavourable. The high speed sparks have insufficient contact time with any element of the fuel–air mixture to provide the energy required for ignition.

## WORKED EXAMPLE 12.4

A fine flammable dust is leaking from a pressurized container at a rate of 2 litres/min into a room of volume 6 $\text{m}^3$ and forming a suspension in the air. The minimum explosible concentration of the dust in air at room temperature is 2.22% by volume. Assuming that the dust is fine enough to settle only very slowly from suspension, (a) what will be the time from the start of the leak before explosion occurs in the room if the air ventilation rate in the room is 4 $\text{m}^3/\text{h}$, and (b) what would be the minimum safe ventilation rate under these circumstances?

### Solution

Mass balance on the dust in the room:

$$\begin{bmatrix} \text{rate of} \\ \text{accumulation} \end{bmatrix} = \begin{bmatrix} \text{rate of flow} \\ \text{into the room} \end{bmatrix} - \begin{bmatrix} \text{rate of flowout} \\ \text{of room with air} \end{bmatrix}$$

assuming constant gas density,

$$V\frac{dC}{dt} = 0.12 - 4C$$

(where 0.12 is the leak rate in $m^3/h$, V is the volume of the room and C is the dust concentration in the room at time t)

Rearranging and integrating with the initial condition, $C = 0$ at $t = 0$,

$$t = -1.5 \ln \left[ \frac{0.12 - 4C}{0.12} \right] \text{hours}$$

Assuming the explosion occurs when the dust concentration reaches the lower flammability limit, 2.22%,

$$\text{time required} = 2.02 \text{ hours.}$$

(b) To ensure safety, the limiting ventilation rate is that which gives a room dust concentration of 2.22% at steady state (i.e. when $dC/dt = 0$). Under this condition,

$$0 = 0.12 - FC_{fL}$$

hence, the minimum ventilation rate, $F = 5.4 \text{ m}^3/h$.

## EXERCISES

**12.1** It is proposed to protect a section of duct used for pneumatically transporting a food product in powder form in air by adding a stream of carbon dioxide. The air flowrate in the present system is $3 \text{ m}^3/s$ and the air carries 2% powder by volume. If the minimum oxygen for combustion (by replacement of oxygen with carbon dioxide) of the powder is 13% by volume, what is the minimum flowrate of carbon dioxide which must be added to ensure safe operation?

(Answer: $1.75 \text{ m}^3/s$)

**12.2** A combustible dust has a lower flammability limit in air at 20°C of 1.2% by volume. A dust extraction system operating at $3 \text{ m}^3/s$ is found to have a dust concentration of 1.5% by volume. What minimum flowrate of additional air must be introduced to ensure safe operation?

(Answer: $0.75 \text{ m}^3/s$)

**12.3** A flammable pharmaceutical powder suspended in air at a concentration within the flammable limits and with an oxygen concentration above the minimum oxygen for combustion flows at 40 m/s through a tube whose wall temperature is greater than the measured ignition temperature of the dust. Give reasons why ignition does not necessarily occur.

**12.4** A fine flammable plastic powder is leaking from a pressurized container at a rate of 0.5 litres/min into another vessel of volume $2 \text{ m}^3$ and forming a suspension in the air in the vessel. The minimum explosible concentration of the dust in air at room temperature is 1.8% by volume. Stating all assumptions, estimate:

(a) the delay from the start of the leak before explosion occurs if there is no ventilation;

(b) the delay from the start of the leak before explosion occurs if the air ventilation rate in the second vessel is $0.5 \text{ m}^3/\text{h}$;

(c) the minimum safe ventilation rate under these circumstances.

(Answer: (a) 1.2 hours; (b) 1.43 hours; (c) $1.67 \text{ m}^3/\text{h}$)

# Notation

| | | |
|---|---|---|
| $a$ | exponent in Equation (11.13) | — |
| $a$ | surface area of particles per unit bed volume | $m^2$ |
| $A$ | cross-sectional area | — |
| $A$ | cyclone body length (Figure 7.5) | $m$ |
| $A$ | surface area of element of fuel–air mixture | $m^2$ |
| $A_f$ | flow area occupied by fluid | $m^2$ |
| $A_p$ | flow area occupied by particles | $m^2$ |
| $A(v)$ | rate of constant for granule attrition | $m^3/s$ |
| $Ar$ | Archimedes number $\left\{ Ar = \dfrac{\rho_f(\rho_p - \rho_f)gx^3}{\mu^2} \right\}$ | — |
| $b$ | exponent in Equation (11.13) | — |
| $B$ | length of the cyclone cone (Figure 7.5) | $m$ |
| $B$ | minimum diameter of circular hopper outlet | $m$ |
| $B_{nuc}(v)$ | rate of granule growth by nucleation | $m^{-6}\,s^{-1}$ |
| $c$ | particle suspension concentration | $kg/m^3$ |
| $C$ | fuel concentration in fuel–air mixture | $m^3 fuel/m^3$ |
| $C$ | length of cylindrical part of cyclone (Figure 7.5) | $m$ |
| $C$ | particle volume fraction | — |
| $C$ | suspension concentration | — |
| $C_B$ | concentration of the reactant in the bubble phase at height $h$ | $mol/m^3$ |
| $C_B$ | suspension concentration in downflow section of thickener | — |
| $C_{BH}$ | concentration of reactant leaving the bubble phase | $mol/m^3$ |
| $C_D$ | drag coefficient (defined by Equation (1.5)) | — |
| $C_F$ | concentration of feed suspension | — |
| $C_{fL}$ | lower flammability limit | $m^3 fuel/m^3$ |
| $C_{fU}$ | upper flammability limit | $m^3 fuel/m^3$ |
| $C_g$ | specific heat capacity of gas | $J/kg\,K$ |
| $C_H$ | concentration of reactant leaving the reactor | $mol/m^3$ |
| $C_L$ | suspension concentration in underflow of thickener | — |

| | | |
|---|---|---|
| $C_p$ | concentration of reactant in the particulate phase | mol/m$^3$ |
| $C_{p_{air}}$ | molar specific heat capacity of air | J/mol K |
| $C_{p_{fuel}}$ | molar specific heat capacity of fuel | J/mol K |
| $C_S$ | sediment concentration | — |
| $C_T$ | suspension concentration in upflow section of thickener | — |
| $C_V$ | suspension concentration in overflow of thickener | — |
| $C_0$ | concentration of reactant at distributor | mol/m$^3$ |
| $C_0$ | initial concentration of suspension | — |
| $C_1$ | constant (Equation (4.22)) | s/m$^6$ |
| $D$ | diameter of bed, bin, cyclone (Figure 7.5), pipe, tube (Equation (4.2)), vessel | — |
| $d_{B_v}$ | equivalent volume diameter of a bubble | m |
| $d_{B_{vmax}}$ | equivalent volume diameter of a bubble (maximum) | m |
| $d_{B_{vs}}$ | equivalent volume diameter of a bubble at the surface | m |
| $D_e$ | equivalent tube diameter (Equation (4.3)) | m |
| $(dp/dt)max$ | maximum rate of pressure rise in dust explosion | bar/s |
| $e$ | coefficient of restitution for granule collision | — |
| $E$ | diameter of solids outlet of cyclone (Figure 7.5) | m |
| $E$ | reaction activation energy in Arrhenius equation (12.2) | s$^{-1}$ |
| $E(S^2)$ | standard error (standard deviation of variance of a sample) | — |
| $E_T$ | total separation efficiency | — |
| $Eu$ | Euler number | — |
| $F$ | cumulative frequency undersize | μm$^{-1}$ |
| $F$ | volumetric flow rate of feed suspension to thickener | — |
| $f^*$ | friction factor for packed bed flow | — |
| $F_D$ | total drag force | N |
| $ff$ | hopper flow factor | — |
| $f_g$ | Fanning friction factor | — |
| $F_{gw}$ | gas-to-wall friction force per unit volume of pipe | Pa/m |
| $F_N$ | number of particles per unit volume of pipe | m$^{-3}$ |
| $f_p$ | solids–wall friction factor, defined by Equation (6.19) | — |
| $F_p$ | form drag force | N |
| $F_{pw}$ | solids-to-wall friction force per unit volume of pipe | Pa/m |
| $F_s$ | drag force due to shear | N |
| $F_v$ | force exerted by particles on the pipe wall per unit volume of pipe | N/m$^3$ |
| $F_{vw}$ | van der Waals force between a sphere and a plane | N |
| $f_w$ | wall factor, $U_D/U\infty$ | — |
| $f(x)$ | differential size distribution dF/dx | m$^{-1}$ or micron$^{-1}$ |
| $g$ | acceleration due to gravity | m/s$^2$ |
| $G$ | solids mass flux $= M_p/A$ | kg/m$^2$ s |
| $G(v)$ | volumetric growth rate constant for coating | m$^3$/s |
| $g(x)$ | weighting function (Equation 3.5) | — |
| $G(x)$ | grade efficiency | — |
| $G(x)$ | linear growth rate constant for coating | m/s |
| $h$ | depth of sampling tube below surface (equation 3.15) | m |

| | | |
|---|---|---|
| $h$ | heat transfer coefficient | W/m² K |
| $h$ | height of interface from base of vessel | m |
| $h$ | thickness of liquid coating on a granule | m |
| $H$ | height of bed | m |
| $H$ | height of powder in bin | m |
| $H$ | height of solids in standpipe | m |
| $h_a$ | measure of roughness of granule surface | m |
| $H_c$ | depth of filter cake | m |
| $H_e$ | equivalent length of flow paths through packed bed (Equation (4.3)) | m |
| $H_{eq}$ | depth of cake equivalent to medium resistance | m |
| $h_{gc}$ | gas convective heat transfer coefficient | W/m² K |
| $h_{gp}$ | gas-to-particle heat transfer coefficient | W/m² K |
| $H_m$ | thickness of filter medium | m |
| $h_{max}$ | maximum bed-to-surface heat transfer coefficient | W/m² K |
| $H_{mf}$ | height of bed at incipient fluidization | m |
| $h_{pc}$ | particle convective heat transfer coefficient | W/m² K |
| $h_r$ | radiative heat transfer coefficient | W/m² K |
| $h_0$ | initial height of interface from base of vessel | m |
| $h_1$ | height defined in Figure 2.9 | m |
| $H(\theta)$ | function given by $H(\theta) = 2.0 + \theta/60$ (Equation (8.5)) | — |
| $j$ | reaction order | — |
| $J$ | length of gas outlet pipe of cyclone (Figure 7.5) | m |
| $k$ | reaction rate constant per unit volume of solids | mol/m³. s |
| $K$ | height of inlet of cyclone (Figure 7.5) | m |
| $K_C$ | interphase mass transfer coefficient per unit bubble volume | s⁻¹ |
| $k_g$ | gas conductivity | W/m K |
| $K_H$ | Hamaker constant (Equation (11.1)) | Nm |
| $K_{ih}^*$ | elutriation constant for size range $x_i$ at height $h$ above distributor | kg/m² s |
| $K_{i\infty}^*$ | elutriation constant for size range $x_i$ above TDH | kg/m² s |
| $K_{St}$ | proportionality constant in Equation (12.6) | bar.m/s |
| $K_1$ | constant (Equation (4.3)) | — |
| $K_2$ | constant (Equation (4.5)) | — |
| $K_3$ | constant (Equation (4.8)) | — |
| $L$ | height above the distributor | m |
| $L$ | length of pipe | m |
| $L$ | volumetric flow rate of underflow suspension from thickener | m³/s |
| $L$ | width of inlet of cyclone (Figure 7.5) | m |
| $L_H$ | length of horizontal pipe | m |
| $L_v$ | length of vertical pipe | m |
| $M$ | mass flow rate of solids feed to separation device | kg/s |
| $M$ | mass of solids in the bed | kg |
| $M_B$ | mass of solids in the bed | kg |
| $m_{B_i}$ | mass fraction of size range $x_i$ in the bed | — |
| $M_c$ | mass flowrate of coarse product at solids discharge | kg/s |
| $M_f$ | mass flowrate of fines product at gas discharge | kg/s |

| | | |
|---|---|---|
| $M_f$ | gas mass flowrate | kg/s |
| MOC | minimum oxygen for combustion | $m^3O_2/m^3$ |
| $M_p$ | solids mass flowrate | kg/s |
| $n$ | exponent in Richardson–Zaki equation in Equation (2.24) | — |
| $n$ | number of cyclones in parallel | — |
| $n$ | number of particles in a sample | — |
| $N$ | diameter of gas outlet of cyclone (Figure 7.5) | m |
| $N$ | number of granules per unit volume the system | — |
| $N$ | number of holes per unit area in the distributor | $m^{-2}$ |
| $N$ | number of samples | — |
| $Nu$ | Nusselt number ($h_{gp}x/k_g$) | — |
| $Nu_{max}$ | Nusselt number corresponding to $h_{max}$ | — |
| $n(v, t)$ | number density of granule volume $v$ at time $t$ | $m^{-6}$ |
| $p$ | pressure | Pa |
| $p$ | proportion of component in a binary mixture | — |
| $p_c$ | capillary pressure (Equation (11.2)) | Pa |
| $Pr$ | Prandtl number ($C_g\mu/k_g$) | — |
| $p_s$ | pressure difference (Equation (4.29)) | Pa |
| $q$ | gas flowrate | $m^3/s$ |
| $Q$ | volume flow rate | $m^3/s$ |
| $Q$ | volume flowrate of gas into bed ($= UA$) | $m^3/s$ |
| $Q$ | volumetric flowrate of suspension to thickener | $m^3/s$ |
| $Q_a$ | rate of heat absorption by element of fuel air mixture | W |
| $Q_f$ | volumetric flow rate of gas/fluid | $m^3/s$ |
| $Q_{in}$ | volumetric flow into granulator | $m^3/s$ |
| $Q_{input}$ | rate of heat input to element of fuel-air mixture | W |
| $Q_{mf}$ | volume flowrate of gas into bed at $U_{mf}$ ($= U_{mf}A$) | $m^3/s$ |
| $Q_{out}$ | volumetric flow out of granulator | $m^3/s$ |
| $Q_p$ | volumetric flow rate of particles/solids | $m^3/s$ |
| $Q_s$ | rate of heat loss to surroundings | W |
| $r$ | radius of curved liquid surface | m |
| $R$ | radius of cyclone body | m |
| $R$ | radius of sphere | m |
| $R$ | universal gas constant | J/mol K |
| R' | drag force per unit projected area of particle | $N/m^2$ |
| $r_c$ | filter cake resistance (Equation (4.17)) | $m^{-2}$ |
| $Re^*$ | Reynolds number for packed bed flow (Equation (4.12)) | — |
| $Re_{mf}$ | Reynolds number at incipient fluidization ($U_{mf}x_{sv}\rho_f/\mu$) | — |
| $Re_p$ | single particle Reynolds number (Equation (1.4)) | — |
| $R_i$ | rate of entrainment of solids in size range $x_i$ | kg/s |
| $r_m$ | filter medium resistance (Equation (4.24)) | $m^{-2}$ |
| $R_p$ | average pore radius | m |
| $S$ | total surface area of population of particles | $m^2$ |
| $S$ | estimate of standard deviation of a mixture composition | — |
| $S_v$ | surface area of particles per unit volume of particles | $m^2/m^3$ |
| $S^2$ | estimate of variance of mixture composition | — |
| $S_B$ | surface area of particles per unit volume of particles | $m^2/m^3$ |

| | | |
|---|---|---|
| $Stk$ | Stokes number (defined by Equation (11.4)) | — |
| $Stk^*$ | critical Stokes number for coalescence | — |
| $Stk_{50}$ | Stokes number for $x_{50}$ | — |
| $t$ | time | s |
| $T$ | reaction temperature | K |
| TDH | transport disengagement height | m |
| $T_g$ | gas temperature | K |
| $T_{ig}$ | ignition temperature | K |
| $T_s$ | solids temperature | K |
| $u$ | granule volume | $m^3$ |
| $U$ | superficial gas velocity $(= Q_f/A)$ | m/s |
| $U_B$ | mean bubble rise velocity | m/s |
| $U_{CH}$ | choking velocity (superficial) | m/s |
| $U_D$ | velocity in a pipe of diameter D | m/s |
| $U_f$ | actual or interstitial gas velocity | m/s |
| $U_{fH}$ | interstitial gas velocity in horizontal pipe | m/s |
| $U_{fs}$ | superficial fluid velocity | m/s |
| $U_{fv}$ | interstitial gas velocity in vertical pipe | m/s |
| $U_i$ | actual interstitial velocity of fluid | m/s |
| $U_{int}$ | interface velocity | m/s |
| $U_m$ | superficial velocity at which $h_{max}$ occurs | m/s |
| $U_{mb}$ | superficial gas velocity at minimum bubbling | m/s |
| $U_{mf}$ | superficial gas velocity at minimum fluidization | m/s |
| $U_{ms}$ | minimum velocity for slugging | m/s |
| $U_p$ | actual particle or solids velocity | m/s |
| $U_{pH}$ | actual solids velocity in horizontal pipe | m/s |
| $U_{ps}$ | superficial particle velocity | m/s |
| $U_{pv}$ | actual solids velocity in vertical pipe | m/s |
| $U_r$ | radial gas velocity | m/s |
| $U_R$ | radial gas velocity at cyclone wall | m/s |
| $U_{rel}$ | relative velocity $(= U_{slip} = U_f = U_p)$ | m/s |
| $|U_{rel}|$ | magnitude of $U_{rel}$ | m/s |
| $U_{relT}$ | relative velocity at terminal freefall | m/s |
| $U_{SALT}$ | saltation velocity (superficial) | m/s |
| $U_{slip}$ | slip velocity $(U_f - U_p)$ | m/s |
| $U_T$ | single particle terminal velocity | m/s |
| $U_{T2.7}$ | single particle terminal velocity for a particle 2.7 times the mean size | m/s |
| $U_{Ti}$ | single particle terminal velocity for particle size $x_i$ | m/s |
| $U_\theta$ | particle tangential velocity at radius $r$ | m/s |
| $U_{\theta R}$ | particle tangential velocity at cyclone wall | m/s |
| $U_\infty$ | velocity in an infinite fluid | m/s |
| $v$ | characteristic gas velocity based on $D$ | m/s |
| $v$ | granule volume | $m^3$ |
| $V$ | granulator volume | $m^3$ |
| $V$ | volume of element of fuel–air mixture | $m^3$ |
| $V$ | volume of filtrate passed | $m^3$ |

| $V$ | volumetric flow rate of overflow suspension from thickener | $m^3/s$ |
| $V_{app}$ | approach velocity of granules | $m/s$ |
| $V_{eq}$ | volume of filtrate that must pass in order to create a cake of thickness $H_{eq}$ | $m^3$ |
| $w$ | average granule volume defined in Equation (11.13) | $m^3$ |
| $w^*$ | critical average granule volume for coalescence | $m^3$ |
| $x$ | granule or particle diameter | $m$ |
| $\bar{x}$ | mean particle diameter | $m$ |
| $x^*$ | critical average granule volume for coalescence | $m$ |
| $\bar{x}_a$ | arithmetic mean diameter (Table 3.4) | $m$ |
| $\bar{x}_{aN}$ | arithmetic mean of number distribution | $m$ |
| $\bar{x}_{aS}$ | arithmetic mean of surface distribution | $m$ |
| $\bar{x}_c$ | cubic mean diameter (Table 3.4) | $m$ |
| $x_{crit}$ | critical particle size for separation (equation 7.20) | $m$ |
| $\bar{x}_g$ | geometric mean diameter (Table 3.4) | $m$ |
| $\bar{x}_h$ | harmonic mean diameter (Table 3.4) | $m$ |
| $\bar{x}_{hV}$ | harmonic mean of volume distribution | $m$ |
| $\bar{x}_{NL}$ | number-length mean | $m$ |
| $\bar{x}_{NS}$ | number-surface mean | $m$ |
| $\bar{x}_p$ | mean sieve size of a powder | $m$ |
| $x_p$ | sieve size | $m$ |
| $\bar{x}_q$ | quadratic mean diameter (Table 3.4) | $m$ |
| $\bar{x}_{qN}$ | quadratic mean of number distribution | $m$ |
| $x_s$ | equivalent surface sphere diameter | $m$ |
| $x_{SV}$ | surface-volume diameter (diameter of a sphere having the same surface/volume ratio as the particle) | $m$ |
| $x_v$ | volume diameter (diameter of a sphere having the same volume as the particle) | $m$ |
| $x_{50}$ | cut size (equiprobable size) | $m$ |
| $y$ | gap between sphere and plane (Equation 11.1) | $m$ |
| $Y$ | factor in Equation (5.30) | — |
| $y_i$ | composition of sample number $i$ | — |
| $z$ | $\log x$ | — |
| $z$ | depth of penetration of liquid into powder mass | $m$ |
| $Z$ | pre-exponential constant in Arrhenius equation (12.2) | $J/mol$ |
| $\alpha_s$ | factor relating linear dimension of particle to its surface area | — |
| $\alpha_v$ | factor relating linear dimension of particle to its volume | — |
| $\alpha$ | significance level | — |
| $\beta$ | coalescence kernel or rate constant | $s^{-1}$ |
| $\beta$ | $(U - U_{mf})/U$ | — |
| $\beta_0$ | coalescence rate constant (Equation (11.12)) | — |
| $\beta_1(u, v)$ | coalescence rate constant (Equation (11.12)) | — |
| $\gamma$ | surface tension | $N/m$ |
| $\delta$ | effective angle of internal friction | (degrees) |
| $\Delta p$ | static pressure drop | $Pa$ |
| $(-\Delta p)$ | pressure drop across bed/cake | $Pa$ |
| $(-\Delta p_c)$ | pressure drop across cake | $Pa$ |

| | | |
|---|---|---|
| $\varepsilon$ | suspension voidage | — |
| $\varepsilon_B$ | volume fraction of bed occupied by bubbles | — |
| $\varepsilon_{CH}$ | suspension voidage at choking | — |
| $\varepsilon_F$ | voidage of feed suspension | — |
| $\varepsilon_H$ | suspension voidage in horizontal pipe | — |
| $\varepsilon_{mf}$ | voidage at $U_{mf}$ | — |
| $\varepsilon_v$ | suspension voidage in vertical pipe | — |
| $\theta$ | dynamic contact angle of liquid with solid | — |
| $\theta$ | semi-included angle of conical hopper | (degrees) |
| $\mu$ | fluid/filtrate viscosity | Pa s |
| $\mu_e$ | effective viscosity of suspension | Pa s |
| $\rho_{ave}$ | effective average suspension density (Equation (2.2)) | kg/m$^3$ |
| $\rho_B$ | bulk density of powder | kg/m$^3$ |
| $\rho_f$ | fluid/filtrate density | kg/m$^3$ |
| $\rho_g$ | gas density | kg/m$^3$ |
| $\rho_{gr}$ | granule density | kg/m$^3$ |
| $\rho_i$ | solids loading of particle size $x_i$ | kg/m$^3$ |
| $\rho_{mair}$ | molar density of air | mol/m$^3$ |
| $\rho_{mfuel}$ | molar density of fuel | mol/m$^3$ |
| $\rho_p$ | particle density | kg/m$^3$ |
| $\rho_T$ | total solids loading of all solids | kg/m$^3$ |
| $\sigma$ | standard deviation of normal distribution | — |
| $\sigma^2$ | true mixture variance | — |
| $\sigma_C$ | compacting/normal stress | kN/m$^2$ |
| $\sigma_C$ | critical stress | kN/m$^2$ |
| $\sigma_D$ | stress developed in the particulate solids | kN/m$^2$ |
| $\sigma_h$ | horizontal pressure exerted by bed of powder | Pa |
| $\sigma_L^2$ | lower limit of $\sigma^2$ for given confidence level | — |
| $\sigma_R^2$ | upper limit of variance of a random mixture | — |
| $\sigma_U^2$ | upper limit of $\sigma^2$ for given confidence level | — |
| $\sigma_v$ | vertical pressure exerted by bed of powder | Pa |
| $\sigma_y$ | unconfined yield stress | kN/m$^2$ |
| $\sigma_z$ | standard deviation of logarithmic distribution | — |
| $\sigma_0^2$ | lower limit of mixture variance | — |
| $\sigma_1$ | major principal stress | kN/m$^2$ |
| $\sigma_2$ | minor principal stress | kN/m$^2$ |
| $\phi$ | volume of cake formed by passage of unit volume of filtrate | m$^3$ |
| $\Phi_A$ | factor in Equation (5.35) | — |
| $\Phi_B$ | factor in Equation (5.32) | — |
| $\Phi_w$ | kinematic angle of wall friction | (degrees) |
| $\chi$ | number of interphase transfer units ($K_C H/U_B$) | — |
| $\chi_a^2$ | lower limit of $\chi^2$ for given confidence level | — |
| $\chi_{1-a}^2$ | upper limit of $\chi^2$ for given confidence level | — |
| $\psi$ | sphericity | — |

# References

Abrahamsen, A. R. and Geldart, D. (1980) *Powder Technology*, **26**, 35.

Adetayo, A. A. and Ennis B. J. (1997) "Unifying approach to modelling granule coales-cence mechanisms", *AIChEJ*, **43**, (4), pp. 927–934.

Allen, T. (1990) *Particle Size Measurement*, 4th Edition, Chapman & Hall, London.

Baeyens, J. and Geldart, D. (1974) "An Investigation into Slugging Fluidized Beds", *Chem. Eng. Sci.*, **29**, 255.

Baskakov, A. P and Suprun, V. M. (1972) "The determination of the convective compo-nent of the coefficient of heat transfer to a gas in a fluidized bed", *Int. Chem. Eng.*, **12**, 53.

Beverloo, W. A., Leniger, H. A. and Van de Velde (1961) *Chem. Eng. Sci.*, **15**, 260–269.

Bodner, S. (1982) *Proceedings of Pneumatech I, International Conference on Pneumatic Transport Technology*, Powder Advisory Centre, London.

Bond, F. C. (1952) The third theory of comminution, *Mining Engineering, Trans.* AIME, **193**, 484–494.

Botterill, J. S. M. (1975) *Fluid Bed Heat Transfer*, Academic Press, London.

Botterill, J. S. M. (1986) Chapter 9: "Fluid bed heat transfer" in *Gas Fluidization Techno-logy*, ed. D. Geldart, Wiley, Chichester.

Capes, C. E. (1980) *Particle Size Enlargement*, Volume 1 of the *Handbook of Powder Technology*, Elsevier, Amsterdam.

Capes, C. E. and Nakamura, L. (1973) "Vertical pneumatic conveying: an experimental study with particles in the intermediate and turbulent flow regimes", *Can. J. Chem. Eng.*, **51**, 31–38.

Carman, P. C. (1937) Fluid flow through granular beds, *Trans. Inst. Chem. Eng.*, **15**, 150–166.

Chhabra, R. P. (1993) *Bubbles, Drops and Particles in Non-Newtonian Fluids*, CRC Press, Boca Raton, Florida.

Clift, R., Grace, J. R. and Weber, M. E. (1978) *Bubbles, Drops and Particles*, Academic Press, London.

Coelho, M. C. and Harnby, N. (1978) "The effect of humidity on the form of water retention in a powder", *Powder Technology*, **20**, p. 197.

Coulson, J. M. and Richardson, J. F. (1991) *Chemical Engineering*, Volume 2: *Particle Technology and Separation Processes*, 4th Edition, Pergamon, Oxford.

Dallavalle, J. M. (1948) *Micromeritics*, 2nd Edition, Pitman, London.

Danckwerts, P. V. (1953) *Research* (London), **6**, 355.

Darcy, H. P. G. (1856) Les fontaines publiques de la ville de Dijon. *Exposition et applications à suivre et des formules à employer dans les questions de distribution d'eau.* Victor Dalamont.

Darton, R. C., La Nauze, R. D., Davidson, J. F. and Harrison, D. (1977) "Bubble growth due to coalescence in fluidised beds", *Trans. Inst. Chem. Engrs.*, **55**, 274.

Davidson, J. F. and Harrison, D. (1971) *Fluidization*, Academic Press, London.

Dixon, G. (1979), The impact of powder properties on dense phase flow, *Proc. Int. Conf. On Pneumatic Transport*, London, UK.

Ennis, B. J. and Litster, J. D. (1997) Section 20: Size enlargement, in Perry's Chemical Engineers' Handbook, 7th Edition, McGraw-Hill, New York.

Ergun, S. (1952) Fluid flow through packed columns, *Chem. Eng. Prog.*, **48**, pp. 89–94.

Evans, I., Pomeroy C. D. and Berenbaum, R. (1961) The compressive strength of coal, *Colliery Engineering*, 75–81, 123–127 and 173–178.

Faxen, H. (1923) *Ark. Mat. Astronom. Fys.*, **17**, (27), 1–28.

Flain, R. J. (1972) "Pneumatic conveying: how the system is matched to the materials", *Process Engineering*, Nov.

Francis, A. W. (1933) Wall effects in falling ball method for viscosity, *Physics*, **4**, 403.

Fryer, C. and Uhlherr, P. H. T. (1980) *Proceedings of CHEMECA '80, the Eighth Australian Chemical Engineering Conference*, Melbourne, Australia, August 24–27.

Geldart, D. (1973) "Types of gas fluidisation" in *Powder Technology*, **7**, 285–292.

Geldart, D. (ed) (1986) *Gas Fluidization Technology*, Chapter 12 by T. M. Knowlton, John Wiley and Sons, Chichester.

Geldart, D. (1990) "Estimation of basic particle properties for use in fluid-particle process calculations", *Powder Technology*, **60**, 1.

Geldart, D. (1992) *Gas Fluidization Short Course*, University of Bradford.

Geldart, D. and Abrahamsen, A. R. (1981) "Fluidization of fine porous powders", *Chem. Eng. Prog. Symp. Ser.*, **77**, No. 205, 160.

Geldart, D., Cullinan, J., Gilvray, D., Georghiades, S. and Pope, D. J. (1979) "The effects of fines on entrainment from gas fluidised beds", *Trans. Inst. Chem. Engrs.*, **57**, 269.

Gilvary, J. J. (1961) "Fracture of brittle solids I, Distribution function for fragment size in single fracture", *J. Appl. Phys.*, **32**, 391–399.

Grace, J. R., Avidan A. A. and Knowlton T. M. (eds.) (1997) *Circulating Fluidized Beds*, chapter by T. M. Knowlton, Blackie Academic and Professional, London.

Griffith, A. A. (1921) *Phil. Trans. R. Soc.*, **221**, 163.

Hamaker, H. C. (1937) "The London-Van der Waals attraction between spherical particles", *Physica*, **4**, 1058.

Harnby, N., Edwards, M. F. and Nienow, A. W. (1992) *Mixing in the Process Industries*, 2nd edition, Butterworth-Heinemann, London.

Hawkins, A. E. (1993) *The shape of powder particle outlines*, Research Studies Press, Wiley, Chichester.

Hinkle, B. L. (1953) PhD Thesis, Georgia Institute of Technology.

Holmes, J. A. (1957) "Contribution to the study of comminution - modified form of Kick's Law", *Trans. Instn. Chem. Engrs.*, **35**, No. 2, 125–141.

Horio, M., Taki, A., Hsieh, Y. S. and Muchi, I. (1980) "Elutriation and particle transport through the freeboard of a gas-solid fluidized bed", in *Fluidization*, eds. J. R. Grace and J. M. Matsen, Engineering Foundation, New York, p. 509.

Hukki, R. T. (1961) Proposal for Solomonic settlement between the theories of von Rittinger, Kick and Bond, *Trans. AIME*, **220**, 403–408.

Inglis, C. E. (1913) "Stress in a plate due to the presence of cracks and sharp corners", *Proc. Instn. Nab. Arch.*

Janssen, H. A. (1895) "Tests on grain pressure silos", *Zeits. d. Vereins Deutsch Ing,* **39**, (35), 1045–1049.

Jenike, A. W. (1964) "Storage and flow of solids", *Bulletin No. 123 of the Utah Engineering Experimental Station,* University of Utah, **53**, (26).

Karra, V. K. and Fuerstenau, D. W. (1977) "The effect of humidity on the trace mixing kinetic in fine powders", *Powder Technology,* **16**, 97.

Kendal, K. (1978) "The impossibility of comminuting small particles by compression", *Nature,* **272**, 710.

Khan, A. R., Richardson, J. F. and Shakiri, K. J. (1978) in Fluidization - Proceedings of the Second Engineering Foundation Conference (Eds. J. F. Davidson and D. L. Keairns), Cambridge University Press, p. 375.

Khan, A. R and Richardson, J. F. (1989) "Fluid-particle interactions and flow characteristics of fluidised beds and settling suspensions of spherical particles", *Chem. Eng. Comm.,* **78**, 111.

Kick, F. (1885) "*Das Gasetz der proportionalen Widerstände und seine Anwendung*", Leipzig.

Klintworth, J. and Marcus, R. D. (1985) "A review of low-velocity pneumatic conveying systems", *Bulk Solids Handling,* **5**, No. 4, pp. 747–753.

Knight, J. B., Jaeger, H. M. and Nagel, S. R. (1993) "Vibration-induced size separation in granular media", *Phys. Rev. Lett.,* **70**, 3728.

Knowlton, T. M. (1986) Chapter 12 on "Solids transfer in fluidized systems", in *Gas Fluidization Technology* - ed. D. Geldart, J. Wiley & Sons, Chichester.

Knowlton, T. M. (1997) "Standpipes and non-mechanical valves", notes for the continuing education course *Gas Fluidized Beds: Design and Operation,* Monash University Department of Chemical Engineering.

Konno, H. and Saito, S. J. (1969) "Pneumatic conveying of solids through straight pipes", *Chem. Eng. Japan,* **2**, 211–217.

Konrad, K., (1986) "Dense phase conveying: a review" *Powder Technology,* **49**, 1–35.

Kozeny, J. (1927) *Sitzb. Akad. Wiss.,* Wien, Math. -naturw. Kl. **136**, (Abt, IIa), 271–306.

Kozeny, J. (1933) *Z. Pfl.-Ernahr. Dung. Bodenk,* **28A**, 54–56.

Kunii, D. and Levenspiel, O. (1969) *Fluidization Engineering,* Wiley, Chichester.

Kunii, D. and Levenspiel, O. (1990) *Fluidization Engineering,* 2nd Edition, Wiley, Chichester.

Lacey, P. M. C. (1954) "Developments in the theory of particulate mixing", *Journal of Applied Chemistry,* **4**, 257.

Leung, L. S. and Jones, P. J. (1978) Paper D1, *Proceedings of Pneumotransport 4 Conference,* BHRA Fluid Engineering.

Liu, L. X. and Litster J. (1993) "Coating mass distribution from a spouted bed seed coater: experimental and modelling studies", *Powder Technology,* **74**, p. 259.

Lunn, G. (1992) *Guide to Dust Explosion, Prevention and Protection,* Part I, *Venting,* I. Chem. E., Rugby.

Mainwaring, N. J. and Reed, A. R. (1987) An appraisal of Dixon's slugging diagram for assessing the dense phase transport potential of bulk solid materials, *Pneumatech 3, Proceedings,* pp. 221–234.

Mills, D. (1990) *Pneumatic Transport Design Guide,* Butterworth, London.

Munroe, H. S. (1888–89) *Trans. AIMME,* **17**, 637–657.

Newitt, D. M. and Conway-Jones J. M. (1958) A contribution to the theory and practice of granulation, *Trans. Inst. Chem. Engrs.,* **36**, 422–442.

Orcutt, J. C., Davidson J. F. and Pigford, R. L. (1962) *CEP. Symp.Ser.,* No. 38, **58**, 1.

Palmer, K. N (1990) Chapter 11: "Explosion and fire hazards of powders", in *Principles of Powder Technology,* ed. M. J. Rhodes, Wiley, Chichester, pp. 299–334.

Perry, R. H. and Green, D. (Eds) (1984) *Perry's Chemical Engineers' Handbook,* 6th Edition, McGraw-Hill, New York, section 20.

Poole, K. R., Taylor, R. F. and Wall, G. P. (1964) "Mixing powders to fine-scale homogeneity: studies of batch mixing", *Trans. Inst, Chem. Eng.*, **42**, T305.

Punwani, D. V., Modi, M. V. and Tarman, P. B. (1976) Paper presented at the *International Powder and Bulk Solids Handling and Processing Conference*, Chicago.

Randolph, A. D. and Larson M. A. (1971) *Theory of Particulate Processes*, Academic Press, London.

Richardson, J. F. and Zaki, W. N. (1954) "Sedimentation and fluidization", *Trans. Inst. Chem. Eng.*, **32**, 35.

Rittinger, R.P. von (1867) *Textbook of Mineral Dressing*, Ernst and Korn, Berlin.

Rizk, F. (1973) Dr-Ing. Dissertation, Technische Hochschule Karlsruhe.

Rumpf, H. (1962) In *Agglomeration*, ed. W. A. Krepper, Wiley, New York, p. 379.

Sastry, K. V. S. (1975) "Similarity of size distribution of agglomerates during their growth by coalescence in granulationof green pelletization", *Int. J. Min. Process.*, **2**, 187.

Sastry, K. V. S. and Fuerstenau D. W. (1970) "Size distribution of agglomerates in coalescing disperse systems", *Ind. Eng. Chem. Fundamentals*, **9**, (1), 145.

Sastry, K. V. S. and Fuerstenau D. W. (1977) In *Agglomeration '77*, ed. K. V. S. Sastry, *AIME*, New York, p. 381.

Sastry, K. V. S. and Loftus K. D. (1989) "A unified approach to the modeling of agglomeration processes", *Proc. 5th Int. Symp. on Agglomeration*, I. Chem. E., Rugby, p. 623.

Schiller, L. and Naumann, A. (1993) "Über die grundlegenden Berechnungen der Schwerkraftaufbereitung", *Z. Ver. deut. Ing.*, **77**, 318.

Schofield, C. (1985), *Guide to Dust Explosion, Prevention and Protection*, Part I, *Venting*, I. Chem. E., Rugby.

Schofield, C. and Abbott J. A. (1988) *Guide to Dust Explosion, Prevention and Protection*, Part II, *Ignition Prevention, Containment, Suppression and Isolation*, I. Chem. E., Rugby.

Stokes, G. G. (1851) "On the effect of the internal friction of fluids on the motion of pendulums", *Trans. Cam. Phil. Soc.*, **9**, 8.

Svarovsky, L. (1981) *Solid–Gas Separation*, Elsevier, Amsterdam.

Svarovsky, L. (1984) "Some notes on the use of gas cyclones for classification of solids", *Proc. First European Symposium on Particle Classification in Gases and Liquids*, Nuremberg, Dechema (May).

Svarovsky, L. (1986) "Solid–gas separation", in *Gas Fluidization Technology*, ed. D. Geldart, Wiley, Chichester, pp. 197–217.

Svarovsky, L. (1990) "Solid–gas separation", in *Principles of Powder Technology*, ed. M. J. Rhodes, Wiley, Chichester, pp. 171–192.

Toomey, R. D. and Johnstone, H. F. (1952) "Gas fluidization of solid particles", *Chem. Eng. Prog.*, **48**, 220–226.

Tsuji, Y. (1983) *Bulk Solids Handling*, **3**, 589–595.

Wen, C. Y. and Yu, Y. H. (1966) "A generalised method for predicting minimum fluidization velocity", *A.I.Ch.E.J.*, **12**, 610.

Werther, J. (1983) "Hydrodynamics and mass transfer between the bubble and emulsion phases in fluidized beds of sand and cracking catalyst", in *Fluidization*, eds. D. Kunii and R. Toei, Engineering Foundation, New York, p. 93.

Williams, J. C. (1990) Chapter 4: "Mixing and segregation in powders" in *Principles of Powder Technology*, ed. M. J. Rhodes, John Wiley & Sons, Chichester.

Wilson, K. C. (1981) "Analysis of slip of particulate mass in a horizontal pipe", *Bulk Solids Handling*, **1**, 295–299.

Yagi, S. and Muchi I. (1952) *Chem. Eng.*, (Japan), **16**, 307.

Zabrodsky, S. S. (1966) *Hydrodynamics and Heat Transfer in Fluidized Beds*, MIT Press, Cambridge, Mass.

Zenz, F. A. (1964) "Conveyability of materials of mixed particle size", *Ind. Eng. Fund.*, **3**, No. 1, pp. 65–75.

Zenz, F. A. (1983) "Particulate solids - the third fluid phase in chemical engineering", *Chem. Engng.*, Nov, p. 61–67.

Zenz, F. A. and Othmer, D. F. (1960) *Fluidization and Fluid-Particle Systems*, Reinhold, New York.

Zenz, F. A. and Weil N. A. (1958) "A theoretical-empirical approach to the mechanism of particle entrainment from fluidized beds", *AIChEJ*, **4**, p. 472.

# Index